Lab Manual

Chemistry
An Atoms First Approach

SECOND EDITION

Steven S. Zumdahl
University of Illinois at Urbana-Champaign

Susan Arena Zumdahl
University of Illinois at Urbana-Champaign

Prepared by

Charles H. Atwood
University of Utah

David R. Thomas
University of Utah

CENGAGE Learning

Australia • Brazil • Mexico • Singapore • United Kingdom • United States

CENGAGE
Learning

WCN: 01-100-101

For product information and technology assistance, contact us at **Cengage Learning Customer & Sales Support, 1-800-354-9706**.

For permission to use material from this text or product, submit all requests online at **www.cengage.com/permissions**
Further permissions questions can be emailed to **permissionrequest@cengage.com**.

ISBN: 978-1-305-39831-3

Cengage Learning
20 Channel Center Street
Boston, MA 02210
USA

Cengage Learning is a leading provider of customized learning solutions with office locations around the globe, including Singapore, the United Kingdom, Australia, Mexico, Brazil, and Japan. Locate your local office at: **www.cengage.com/global**.

Cengage Learning products are represented in Canada by Nelson Education, Ltd.

To learn more about Cengage Learning Solutions, visit **www.cengage.com**.

Purchase any of our products at your local college store o at our preferred online store **www.cengagebrain.com**.

Printed in the United States of America
1 2 3 4 5 20 19 18 17 16

Lab Manual to Accompany
Chemistry: An Atoms First Approach
Table of Contents

Experiment 1
Making and Recording Measurements in the Lab

Objectives

- Become familiar with common laboratory equipment and its proper use
- Learn how to report measurements with the appropriate level of uncertainty
- Determine the density of both solids and liquids

Topics Introduced in this Lab

- Dimensional analysis
- Mole
- Uncertainty and significant figures
- Temperature
- Density

Related Zumdahl Sections

- Review Chapter Sections 1 - 7
- Review Chapter Section 10

Safety Considerations

- Appropriate lab attire must be worn at all times when working in the lab.
- Dispose of all liquids as indicted by your insttructor.
- Solids (metals and pennies) should be returned where indicated by your instructor
- Do not ingest any of the chemicals used in this experiment

Introduction

As a chemist, you will be constantly required to make observations. Most of these observations will be quantitative in nature and are referred to as measurements. Just as grammar is essential in correctly translating ideas in written language (i.e., "Let's eat dad!" versus "Let's eat, dad!"), the ability to accurately report measurements is essential in correctly translating information in the field of science.

In this experiment, you will use some of the common laboratory equipment available to chemists to measure various quantities, including length, mass, temperature, and volume. You will practice reporting these values to the correct number of significant figures, thereby correctly indicating the appropriate level of uncertainty associated with the measurements. Additionally, you will utilize these measurements to calculate the densities of several solids and liquids.

Procedure

Part A – Measuring Length and Using Dimensional Analysis

1. Obtain a penny from your instructor that is post 1982.
2. Using a ruler, you will measure the dimensions of the penny in cm. Before you do this, examine your ruler. What is the smallest division indicated on the ruler?Keeping in mind that you should always record one digit past the smallest division on your measuring device, to what decimal place should you record your length measurements? *Be sure to always include units when recording measurements.*
3. Measure and record the diameter and the thickness of the penny (in cm) to the correct number of significant figures.
4. Steps 5 and 6 can be completed outside of lab. Proceed to Part B. Keep your penny, as you will need it later in the experiment.
5. Assume that you had a mole of pennies. How much money in dollars would you have? If you stacked the mole of pennies one on top of the other, how many feet tall would your stack be? How tall would this be in miles? (2.54 cm = 1 in (exactly), 5280 ft = 1 mi (exactly))
6. The state of Utah has a land area of approximately 84,899 mi^2. Suppose you placed your one mole of pennies side by side until the entire state of Utah was covered. Then, you began adding layers on top of the first until you ran out of pennies. How many full layers would there be when you ran out of pennies? How tall (only counting the full layers) would the stack be in meters and in miles?

Part B – Measuring Temperature in the Lab

7. Fill a 250-mL beaker about half full with water.
8. Using a thermometer, measure and record the temperature of the water in degrees Celsius. What is the smallest division indicated on the thermometer? Keeping in mind that you should always record one digit past the smallest division on your measuring device, to what decimal place should you record your temperature measurement?
9. Step 10 can be completed outside of lab. Proceed to Part C.
10. Convert your measured temperature in degrees Celsius to degrees Fahrenheit and to Kelvin. Be sure to show all work.

Part C – Measuring Volumes in the Lab

11. Fill a 50- or 100-mL beaker with water to the 50 mL line exactly. The volume on a graduated cylinder or beaker should always be read at the bottom of the meniscus for water.
12. Being sure to transfer all of the water, carefully pour the contents of the beaker into a 100-mL graduated cylinder.
13. Record the volume of water in mL as measured using the graduated cylinder. What is the smallest division indicated on the graduated cylinder? Keeping in mind that you should always record one digit past the smallest division on your

measuring device, to what decimal place should you record your volume measurement?

14. Graduated cylinders are designed to measure volumes of liquids accurately. Keeping that in mind, and comparing the volume of water measured using the beaker to that measured with the graduated cylinder, which is the best device for accurately measuring volume? Explain your answer.

Part D – Measuring Mass in the Lab

15. Using the same penny that you used to make length measurements earlier, proceed to the balance area.
16. Analytical balances are precision instruments and even something as minor as air currents can affect them. After making sure that the balance is clean and all doors are closed, press the TARE or ZERO button and wait for the balance to stabilize at 0.
17. After the balance stabilizes, open a door and gently place your penny on the balance's weigh plate. Close the door and wait for the balance to stabilize before recording the mass shown in grams.
18. After recording the mass of the penny, and with the penny still on the balance, hit the TARE or ZERO button again. Note that this "zeroes" the balance, no longer accounting for the mass of the penny but still allowing the penny to remain on the balance. In a similar fashion, containers used to hold chemicals can be "zeroed" before a chemical is weighed into the container.
    (*Note: It is always best to use the same balance throughout an experiment since slight variations exist between balances.)
19. Remove your penny from the balance (being sure to close all balance doors) and return the penny to your instructor.

Part E – Combining Mass and Volume Measurements to Calculate Densities

20. Steps 21-23 can be completed outside of lab. Proceed to Step 24.
21. Using the dimensions of the penny determined in Part A, calculate the volume of the penny in cm^3. ($V_{cylinder} = \pi r^2 h$)
22. Using the volume you just calculated and the mass of the penny determined in Part D, calculate the density of the penny.
23. Pennies are a mixture of copper and zinc. The density of pure copper is 8.94 g/cm^3 and the density of pure zinc is 7.13 g/cm^3. Based on your calculated density, which element is used as the major component in the production of post-1982 pennies? If you are unable to make this determination based on your data, explain why.
24. Using what you have learned about measuring masses and volumes in the lab, come up with a way to determine the density of rubbing alcohol. Be sure to record all relevant data (masses and volumes) and show your calculation in the lab report.
25. Perform a second trial to determine the density of rubbing alcohol. Is the density determined in the second trial exactly the same as it was in the first? What does this tell you about the usefulness of performing multiple trials?

26. Using the same technique, determine the density of vegetable oil. Perform at least 2 trials.
27. Obtain an unknown salt solution from your instructor. Record the unknown solution's number in your lab report. Determine the density of the salt solution. Perform at least 2 trials.
28. Obtain an unknown metal from your instructor. Record the unknown metal's number in your lab report. Determine the density of the metal. (Remember that you must determine both the mass of the metal and its volume to calculate a density). Perform at least 2 trials.
29. The metal is one of the following:
    - Aluminum
    - Copper
    - Iron
    - Magnesium

    Using an outside resource, such as your textbook, the internet, the library, etc., along with your calculated density, determine the identity of the unknown metal. Be sure to reference your resource in your report.

Experiment 1 Pre-Lab (Show ALL Work and ALWAYS include units)

Use **Figure 1** below to answer questions 1 and 2.

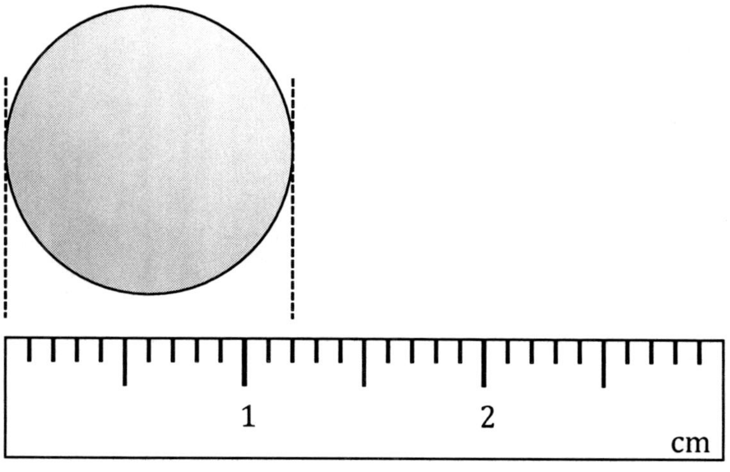

Figure 1. Measuring a Metal Sphere with a Ruler

1. You measure the diameter of a metal sphere using a ruler as shown above. What is the smallest division on the ruler?

   a. 1 cm   b. 0.1 cm   c. 0.01 cm   d. 0.001 cm

2. Keeping in mind that you should always report one digit past the smallest division on your measuring device (all of the certain digits plus the first uncertain digit), report the diameter of the metal sphere to the correct number of significant figures.

   ____________ cm

3. You take the sphere and put it into a graduated cylinder containing water. You examine the volume of water before and after adding the sphere to determine its volume (See **Figure 2** below).

   What is the volume of the metal sphere as determined by this method? Report the correct number of significant figures in your answer.

   ____________ mL

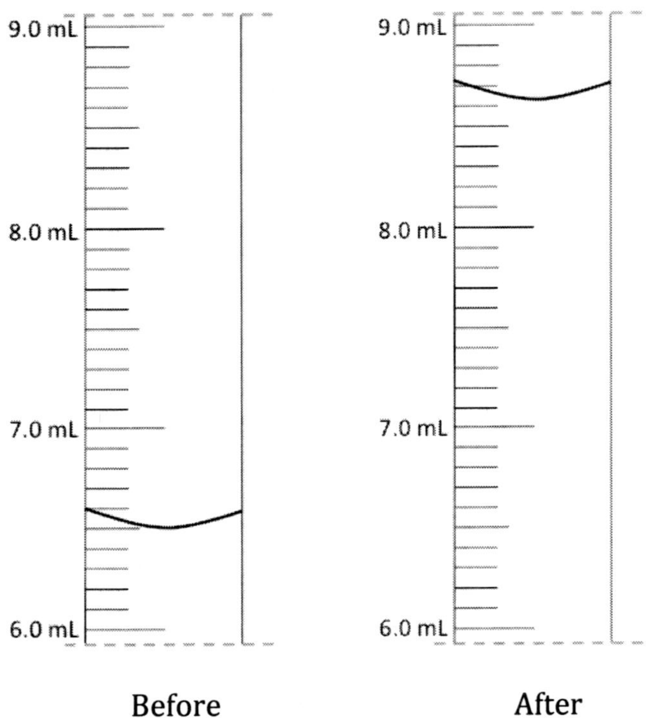

Figure 2. Water Level in Graduated Cylinder Before and After Addition of the Metal Sphere

4. Using a balance, you determine that the metal sphere has a mass of 16.506 g. What is the density of the sphere in g/mL?

_________ g/mL

5. Assume that the sphere is made of a pure metal. Using the density and the information provided in **Table 1** below, identify the metal.

Metal	Density (g/mL)
Iron	7.86
Lead	11.34
Zinc	7.13

The identity of the metal is ____________________.

Experiment 1 Lab Report (Show ALL Work and ALWAYS include units)

Part A - Measuring Length and Using Dimensional Analysis

1. Measuring in cm, to what decimal place is the smallest division on the ruler?

2. Based on your answer to the above question, to what decimal place should length measurements made using the ruler be recorded? Why?

3. Diameter of the penny (in cm): ___________

4. Thickness of the penny (in cm): ___________

5. Assume that you had a mole of pennies. How much money in dollars would you have?

6. If you stacked the mole of pennies one on top of the other, how many feet tall would your stack be? How tall would this be in miles? (2.54 cm = 1 in, 5280 ft = 1 mi) (Both conversions are exact.)

7. The state of Utah has a land area of approximately 84,899 mi^2. Suppose you placed your one mole of pennies side by side until the entire state of Utah was covered. Then, you began adding layers on top of the first until you ran out of pennies. How many full layers would there be when you ran out of pennies. How tall (only counting the full layers) would the stack be in meters and in miles?

Part B – Measuring temperature in the lab

8. Measuring in °C, to what decimal place is the smallest division on the thermometer?

9. Based on your answer to the above question, to what decimal place should temperature measurements made using the thermometer be recorded? Why?

10. Temperature of water (in °C): ___________

11. Calculate the temperature of the water in °F:

Temperature of water (in °F): ___________

12. Calculate the temperature of the water in K:

Temperature of water (in K): ___________

Part C – Measuring Volumes in the Lab

13. Measuring in mL, to what decimal place is the smallest division on the graduated cylinder?

14. Based on your answer to the above question, to what decimal place should volume measurements made using the graduated cylinder be recorded? Why?

15. Volume of the water (in mL): ___________

16. Graduated cylinders are designed to measure volumes of liquids accurately. Keeping that in mind, and comparing the volume of water measured using the beaker to that measured with the graduated cylinder, which is the best device for accurately measuring volume? Explain your answer.

Part D – Measuring Mass in the Lab

17. Mass of the penny (in grams): ____________

Part E - Using Mass and Volume Measurements to Calculate Densities

18. Calculate the volume of the penny:

19. Volume of the penny (in cm^3): ____________

20. Calculate the density of the penny:

21. Density of the penny (in g/cm^3): ____________

22. Pennies are a mixture of copper and zinc. The density of pure copper is 8.94 g/cm^3 and the density of pure zinc is 7.13 g/cm^3. Based on your calculated density, which element is used as the major component in the production of post-1982 pennies? If you are unable to make this determination based on your data, explain why.

23. Data and calculation to determine the density of rubbing alcohol (Trial 1):

Density of rubbing alcohol Trial 1 (in g/mL): __________

24. Data and calculation to determine the density of rubbing alcohol (Trial 2):

Density of rubbing alcohol Trial 2 (in g/mL): __________

25. Is the density determined in the second trial exactly the same as it was in the first? What does this tell you about the usefulness of performing multiple trials?

26. Average density of rubbing alcohol from 2 trials (in g/mL): __________

27. Data and calculation to determine the density of vegetable oil (Trial 1):

Density of vegetable oil Trial 1 (in g/mL): __________

28. Data and calculation to determine the density of vegetable oil (Trial 2):

Density of vegetable oil Trial 2 (in g/mL): __________

29. Average density of vegetable oil (in g/mL): __________

30. Unknown salt solution's number: __________

31. Data and calculation to determine the density of the unknown salt solution (Trial 1):

Density of salt solution Trial 1 (in g/mL): __________

32. Data and calculation to determine the density of unknown salt solution (Trial 2):

Density of salt solution Trial 2 (in g/mL): __________

33. Average density of salt solution (in g/mL): __________

34. Unknown metal's number: __________

35. Data and calculation to determine the density of metal (Trial 1):

Density of metal Trial 1 (in g/mL): __________

36. Data and calculation to determine the density of metal (Trial 2):

Density of metal Trial 2 (in g/mL): __________

37. Average density of metal (in g/mL): __________

38. Identity of metal: ___________________________

Reference:

Experiment 2
The Scientific Method in Action

Objective
- Apply the scientific method to understand unfamiliar phenomena

Topics Introduced in this Lab
- Classification of matter
- Energy
- Thinking "microscopically"
- Scientific method

Related Zumdahl Sections
- Review Chapter Sections 8 and 9
- Chapter 1 Sections 1 and 2

Safety Considerations
- Appropriate lab attire must be worn at all times when working in the lab.
- Do not ingest any of the chemicals used in this experiment

Introduction
At its core, science is a pursuit of understanding. As scientists we attempt to explain how the world works around us one phenomenon at a time. The scientific method provides an organized framework for the construction of this understanding and consists of three primary steps as outlined in **Figure 1**.

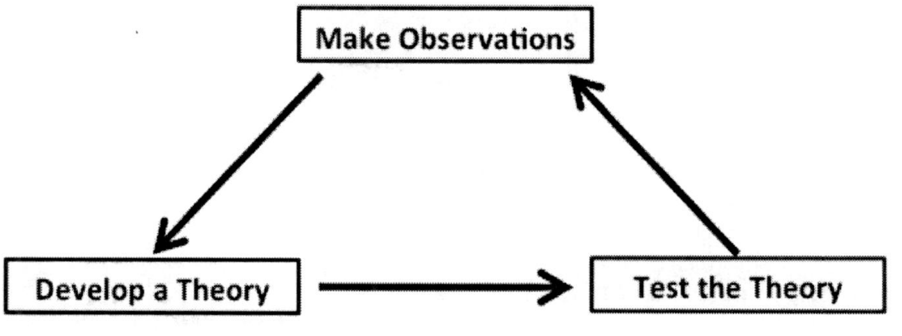

Figure 1. Three Primary Steps of the Scientific Method

The scientific method begins with making observations. Then, if we do not understand what we observe, an attempt is made to explain the observations by creating a theory or a model that provides the answer to "why" something is occurring. Then the question becomes, "How do we know if our theory is correct?" To find out, we perform carefully chosen experiments designed to test our theory. It is important to note that while tests can disprove a theory, they can NEVER prove a theory. Instead, the results of a test can only support a theory. However, as we accumulate more and more test results that support a theory, the more confidence we have in that theory and its ability to explain the observations associated with a given phenomenon.

In this experiment, you will examine several different phenomena. You will utilize the scientific method to try and explain the "Why?" behind your observations. In doing so, you will have to think about the different phases of matter from a microscopic perspective. In other words, what microscopic events are causing the phenomena observed on the macroscopic level?

Procedure

<u>Part A – Heating and Cooling Sodium Chloride</u>

1. Obtain a crucible.
2. Add approximately 1 gram of sodium chloride (table salt) to the crucible.
3. Set up a ring stand with an iron ring, clay triangle, and Bunsen burner as shown in **Figure 2** below. However, do not place the burner under the crucible yet.

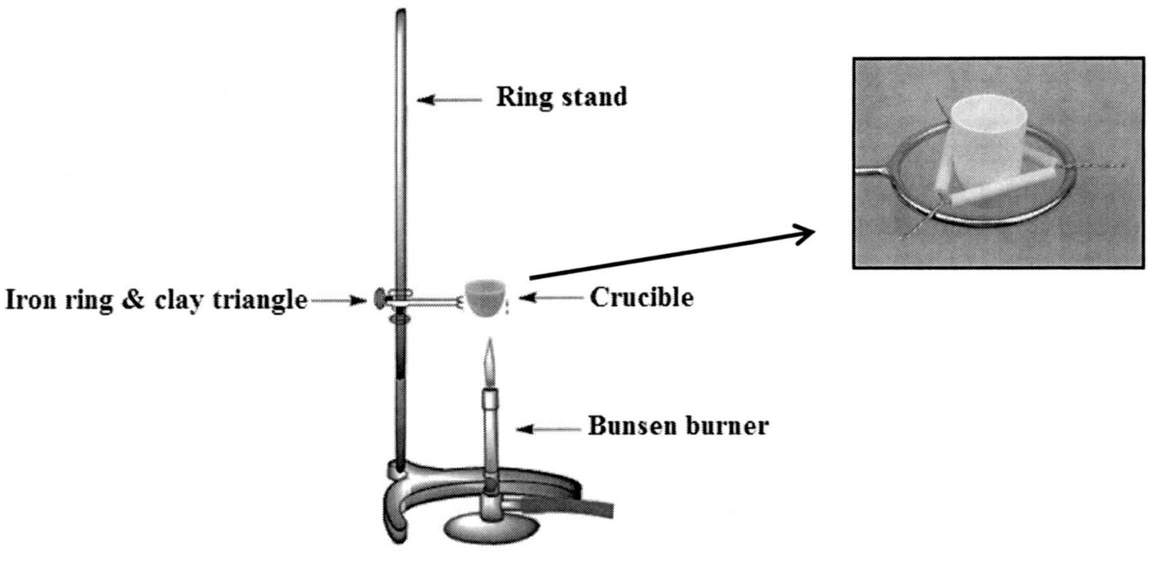

Figure 2. Experimental Setup for Part A

4. Place the crucible in the clay triangle as shown. Turn on the Bunsen burner and adjust the flame using the air intake and gas flow valves until a single blue cone is seen. Do not place the burner under the crucible yet.
5. Lower the ring on the ring stand until the bottom of the crucible is about an inch above where the tip of the blue cone will be when the burner is placed under the crucible. Place a lid (not pictured) on the crucible using crucible tongs, making sure to leave the lid slightly ajar with the opening pointing away from yourself and other students. Now, place the Bunsen burner under the crucible and heat using this "lower setting" for 5 min. *See note below.

*Note: Once the Bunsen burner has been under the crucible and ring for any length of time, the whole experimental setup will get VERY hot. Do not try and adjust the ring height after the initial placement of the crucible. Instead, small adjustments in flame height can be made (if needed) by adjusting the air intake and gas flow valves on the bottom of the burner.

6. After 5 minutes, use the air intake and gas flow adjustments on the Bunsen burner to adjust the flame until two blue cones are clearly seen and the tip of the inner cone is touching the bottom of the crucible. The tip of the inner blue cone is the hottest part of the flame.
7. Leave the lid on the crucible for an additional 5 minutes while heating using this "high setting". After that time, remove the crucible lid using crucible tongs. If a change has not already occurred, watch the sodium chloride until you observe a change. However, do NOT put your face directly over the crucible. Describe what happens to the sodium chloride states of matter over time. Is the change you observe a physical or chemical change? Heat is a form of energy. Keeping that in mind, how would you describe the amount of energy it took to cause the observed change ?
8. Turn off the Bunsen burner and allow the crucible to cool completely. During the cooling process does the sodium chloride change states of matter? Describe what happens to the sodium chloride in the crucible over time. Is the change you observe a physical or chemical change? Since the sodium chloride is now cool, energy in the form of heat has left the sodium chloride. What does this tell you about the relative energies of the two states of matter you observed with the sodium chloride in this experiment? Rinse out the crucible in the sink until the entire residue is removed.

<u>Part B – Interaction of Sodium Chloride with Water</u>

9. Add approximately 1 gram of sodium chloride to approximately 100 mL of water in a beaker and stir with a stirring rod until you observe a change. Record your observations. Pour the contents of the beaker down the drain.

<u>Part C – Cooking with Crisco</u>

10. Obtain a scoop of Crisco shortening (about 30 mL) and place it into a 50-mL beaker.
11. Place the 50-mL beaker on a hotplate and turn the temperature to a medium setting. Watch until the Crisco completely melts. Is this a physical or a chemical change?
12. After all the Crisco has melted, carefully remove the beaker from the hotplate and place it on the lab bench to cool. As the Crisco cools, note which part of the liquid solidifies first. What does this tell you about the two physical states of the Crisco relative to each other? After the Crisco is completely cool, ask your instructor if they want to re-collect the Crisco or if you should throw it in the trash. Use soap and water to clean out any remaining residue in your beaker and any other equipment used with the Crisco.

Part D – Water/Ice

13. Put approximately 300 mL of water in a 400-mL beaker.
14. Using a 50-mL beaker as a scoop, add 1 scoop of ice to the beaker containing water. Does the ice sink or float? What does this tell you about the two physical states of water relative to each other? How does this differ from the Crisco? Pour the contents of the beaker down the drain.

Part E – Digging Deeper

15. Add approximately 200 mL of ice to 50-mL of deionized or distilled water in a 250-mL beaker. Monitor the temperature of the ice/water bath with a thermometer. Once the temperature stops decreasing, record the lowest temperature reached by the ice/water bath. Pour the contents of the beaker down the drain.
16. Repeat the same procedure as in step 15, except this time dissolve approximately 5 grams of sodium chloride in the 50 mL of water before adding the ice. Record the lowest temperature reached. Pour the contents of the beaker down the drain.
17. Repeat the same procedure as in step 16, but use 10 grams of sodium chloride instead of 5. Record the lowest temperature reached. Pour the contents of the beaker down the drain.

Experiment 2 Pre-Lab (Show ALL Work and ALWAYS include units)

1. Which of the following BEST describes a theory/model? (Circle the letter of the correct answer.)

   a. A scientific statement of an observable fact
   b. An explanation of a phenomenon that has limitations and often exceptions
   c. An explanation of a phenomenon that is infallible
   d. An explanation of a phenomenon that can be proven through experiments

2. If solid substance A sinks when placed into a liquid substance B, what is true about substance A relative to substance B?

3. If the substances A and B in question #2 above are actually two states of the same substance, indicate pictorially (in the spaces provided) what the microscopic particles that make up the substance look like in the two different states. Use your answer to question #2 above to help you. Use circles to represent particles.

Solid state

Liquid state

4. Think about the pictures you drew above in question #3. Write down a very basic theory (1–2 sentences) explaining why you think the particles illustrated do not simply spread infinitely far apart, but instead stay together.

Experiment 2 Lab Report (Show ALL Work and ALWAYS include units)

Part A – Heating and Cooling Sodium Chloride

1. Describe what happens to the sodium chloride in the crucible upon heating over time using the states of matter. Is the change you observe a physical or chemical change?

2. Heat is a form of energy. Keeping that in mind, how would you describe the amount of energy it took to cause the change that you observed?

3. During the cooling process does the sodium chloride change? Describe what happens to the sodium chloride states of matter over time. Is the change you observe a physical or chemical change?

4. Once the sodium chloride has cooled, energy in the form of heat has left the sodium chloride. What does this tell you about the relative energies of the two states of matter you observed with the sodium chloride in this experiment?

5. Draw a pictorial representation of the microscopic view of the two states observed for the sodium chloride in the spaces provided below based on what you have learned about the states of matter. Use circles to represent individual sodium chloride particles.

State observed when cool

State observed when heated

6. Using the pictorial representations you gave above in question #5, along with your answer to question #4, present a theory that explains your observations from heating and cooling sodium chloride. Be sure to theorize about what is happening on the microscopic level. Use the words "particles," "energy," and "attractive forces" in your theory.

   Theory:

Part B – Interaction of Sodium Chloride with Water

7. Record your observations after adding the sodium chloride to water.

8. Draw a pictorial representation of the microscopic view of the dissolved sodium chloride in the space given below. Use circles to represent individual sodium chloride particles and triangles to represent water particles.

Microscopic view of dissolved sodium chloride

9. Compare the picture you drew in question #8 with the two pictures you drew in question #5. Which picture more closely resembles the one you just drew, the sodium chloride when cool or when heated? Explain any similarities between the pictures that are most closely related.

10. Using what you've learned, what is the main difference between melting and dissolving? If heat is just a form of energy, then what must be true about the process of dissolving something?

11. Modify/expand your earlier theory developed in question #6 to include both what happens when sodium chloride is heated/cooled AND when sodium chloride is dissolved. Assume that all matter behaves similarly to the sodium chloride in these two basic processes and phrase your theory with respect to all matter, not just sodium chloride.

Theory:

Part C – Cooking with Crisco

12. In step #12 of the procedure, which part of the liquid solidifies first (top, bottom, or all at once)? What does this tell you about the two physical states of the Crisco relative to each other? Does this support or disprove the theory you proposed in question #11 above? Explain. (Be sure to include energy as part of your discussion).

13. Crisco melts at relatively low temperatures (<100 °C). Sodium chloride melts at around 800 °C! Using your theory, attempt to explain why there is such a large difference between the two melting temperatures.

Part D – Water/Ice

14. After adding the ice to the water, did the ice sink or float? What does this tell us about the two physical states of water relative to each other? How does this differ from the Crisco?

15. Do the observations you made of the ice/water mixture support your theory? Explain.

16. Models/theories are inherently limited. The more simplistic a model is, the more exceptions you will find to that theory. Assume that your theory is correct and that the observations in question 14 represent an exception to that theory. Based on your observations, draw a pictorial representation of the microscopic view of these two states of H_2O in the spaces provided below. Use circles to represent individual water particles.

Microscopic view of ice

Microscopic view of liquid water

Part E – Digging deeper

17. Record the lowest temperatures reached by the three baths in steps 15, 16, and 17.

    Ice/Water bath: ________ °C

    Ice/Water/Salt (5 g) bath: ________ °C

    Ice/Water/Salt (10 g) bath: ________ °C

18. How does the lowest temperature of the bath containing 5 grams of sodium chloride compare to the lowest temperature of the bath containing no salt? Building upon your earlier theory, devise another theory to explain why adding 5 grams of sodium chloride to the ice/water bath had the observed effect. (Think on the microscopic level.) Feel free to draw pictures to aid in your explanation.

19. How does the lowest temperature of the bath containing 10 grams of sodium chloride compare to the lowest temperature of the bath containing only 5 grams of sodium chloride? Does this result support or disprove the theory you proposed in question #18? Explain.

Experiment 3
Flame Tests –
Colors Emitted by Compounds When Heated

Objectives

- See one of the clues to atomic structure
- Begin to understand the basis of emission spectra

Topics Introduced in this Lab

- Flame tests
- Emission spectra

Related Zumdahl sections

- Chapter 1 Sections 6 and 7
- Chapter 2 Sections 1, 3, and 4

Safety Considerations

- Methanol is highly flammable and readily absorbed through skin. Do not use near an open flame. Do not get methanol on your skin. Methanol can also lead to blindness if ingested. Handle methanol with care.
- 6 *M* HCl is a strong and concentrated acid. If you get any on your skin wash with copious amounts of water. Handle carefully. Dispose of HCl as indicated by your instructor.
- Compounds used in this experiment may irritate your eyes or lungs if they come into contact with either.
- When heated in a Bunsen burner, nichrome wire reaches temperatures of 500 to 600 °C. After being removed from the flame, the wire does not look as if it is hot but it is and will remain so for several minutes. Do not touch the nichrome wire loop after it is removed from the flame.
- Do not ingest any of these compounds.
- Wear safety glasses and appropriate lab wear at all times while in the lab.

Introduction

In the late 1800's and early 1900's scientists were intrigued by the fact that some chemical compounds produced colored flames when burned. These scientists realized that the colors produced by the flames held some fundamental knowledge of nature. Consequently, these flame tests, as they were called, became subjects of great interest. Indeed, the existence of some elements was inferred based upon their flame tests.

In this experiment we will look at the flame tests of a few chemical compounds. As you perform the experiment, note if the color produced by the flame test corresponds with the color of the compound. Also, note if the flame test color is associated with the positive ion (cation) or the negative ion (anion). Then we shall briefly examine if there is a scientifically sound explanation for the color formation.

Procedure

Part A – Flame Tests

1. Two methods are commonly used to produce flame tests.
    a. One is to place the chemical compound into a petri dish and then spray a mixture of methanol and water on top of the compound. The mixture is then lit with a match or lighter. After the methanol's blue-colored flame heats the chemical compound sufficiently, the flame color will change to that associated with the compound, **Figure 1**.
    b. The second method is to make 0.1 $M(aq)$ solutions of each compound. Then, dip a loop of nichrome wire into the solution. Place the wet nichrome wire into a Bunsen burner flame to produce the flame test. Be sure to clean the nichrome wire with 6 M $HCl(aq)$ after each flame test and before starting another flame test, **Figure 1**.
    c. Ask your instructor which method will be used in your lab.
2. Obtain either 0.1 $M(aq)$ solutions or 1.0-g samples of solid $NaClO_4$, NaCl, $LiCrO_4$, LiCl, KBr, KCl, $CaCl_2$, $CaSO_4$, $Sr(NO_3)_2$, $SrCl_2$, $Cu(NO_3)_2$, and $CuCl_2$.
3. If using method 1b, skip to step 4. If using method 1a, place 0.5- to 1.0-g samples of each compound in separate petri dishes. Using a wash bottle, squirt approximately 5 mL of water-methanol mixture into each petri dish. Using a match or lighter, ignite the methanol in the petri dish and then observe and record the color of each flame. Remove the bottle of methanol from your work area before igniting the methnol-water mixture.
4. If using method 1b, pour approximately 10 mL of each solution in separate clean 50-mL beakers. Ignite a Bunsen burner. Immerse the nichrome wire loop into a solution, then place the loop into the Bunsen burner flame. Observe and record the flame color. Repeat if necessary. Clean the nichrome wire with 6 M HCl then wash with distilled or deionized water. Immerse it into the Bunsen burner flame to burn off any remaining residue before testing a different solution. Observe and record flame colors for each solution.

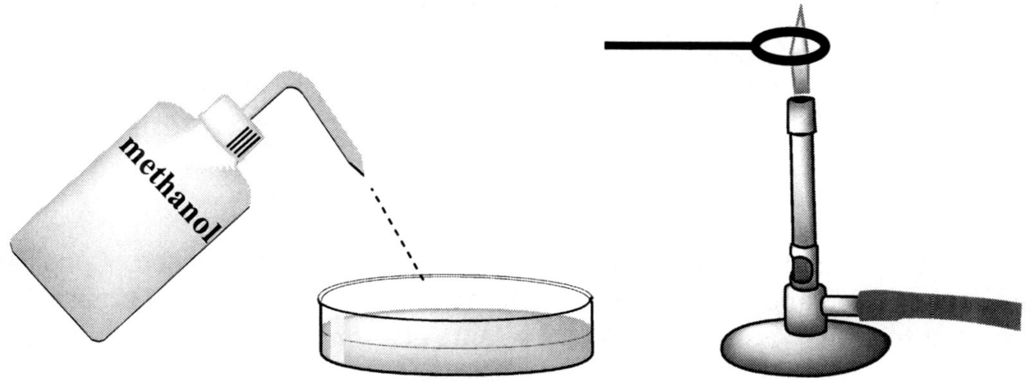

Figure 1. Experimental Setup for Methods 1a and 1 b

Experiment 3 Pre-Lab (Show ALL Work and ALWAYS include units)

1. A purple-colored compound composed of positive ions of element X and negative ions of element Y produces a purple-colored flame test. A white-colored compound composed of positive ions of element X and negative ions of element Z also produces a purple-colored flame test. From this experiment, what can we conclude regarding the relationship of compound color and flame test color?

2. A purple-colored compound composed of positive ions of element X and negative ions of element Y produces a purple-colored flame test. A white-colored compound composed of positive ions of element X and negative ions of element Z also produces a purple-colored flame test. From this experiment what can we conclude regarding the relationship of flame test and ions present in a compound? Assume that X is the same element used in question 1.

3. If we perform a flame test on a green-colored compound composed of positive ions of element X and negative ions of element Q, what color do we expect the flame test to emit? Assume that X is the same element used in questions 1 and 2.

4. If we perform a flame test on a yellow-colored compound composed of positive ions of element R and negative ions of element Y, what color do we expect the flame test to emit? Assume that Y is the same element used in questions 1 and 2

Experiment 3 Lab Report (Show ALL Work and ALWAYS include units)

Part A – Flame Tests

1. Color of $NaClO_4$ flame test __________
2. Color of NaCl flame test __________
3. Color of $LiCrO_4$ flame test __________
4. Color of LiCl flame test __________
5. Color of KBr flame test __________
6. Color of KCl flame test __________
7. Color of $CaCl_2$ flame test __________
8. Color of $CaSO_4$ flame test __________
9. Color of $Sr(NO_3)_2$ flame test __________
10. Color of $SrCl_2$ flame test __________
11. Color of $Cu(NO_3)_2$ flame test __________
12. Color of $CuCl_2$ flame test __________

13. From the flame tests of $NaClO_4$ (a yellow-colored compound) and NaCl (a white-colored compound), what conclusion can be drawn about the correlation of a compound's color and its flame test?

14. From the flame tests of $NaClO_4$ and NaCl, what conclusion can be drawn about the correlation of the ions in a compound and the compound's flame test?

15. From the flame tests of $LiClO_4$ and LiCl, what conclusion can be drawn about the correlation of the ions in a compound and the compound's flame test?

16. From the flame tests of KBr and KCl, what conclusion can be drawn about the correlation of the ions in a compound and the compounds flame test?

17. In the remaining flame tests did you find evidence that confirmed or contradicted what you determined in the $NaClO_4$, NaCl, $LiClO_4$, LiCl, KBr, and KCl flame tests? Based on this evidence, what is the overall conclusion that can be drawn about the species responsible for the colors seen in flame tests?

18. Using the Bohr model representation shown below, give your best explanation of how the flame test colors are generated in LiCl and NaCl.

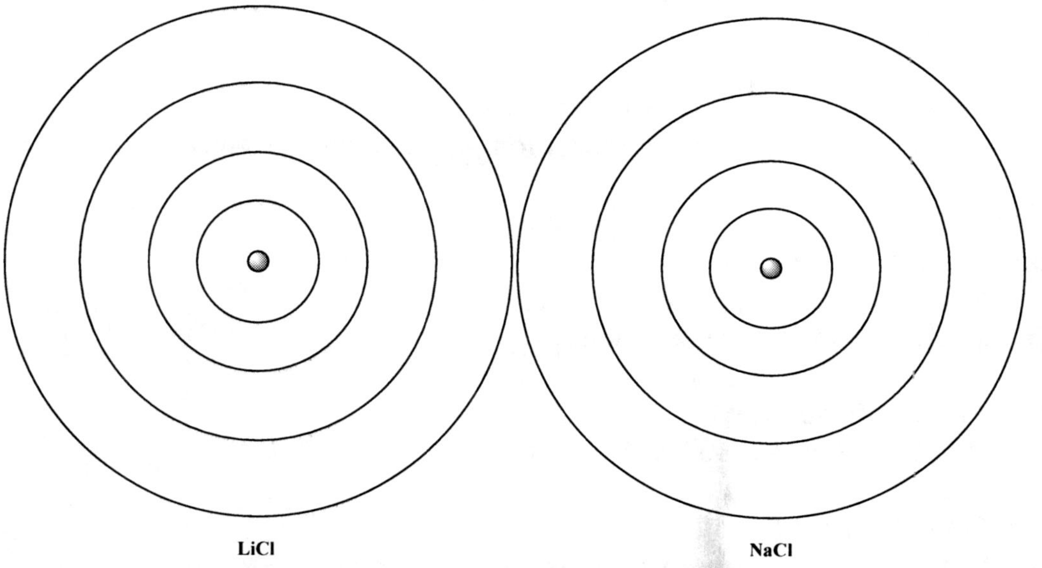

Experiment 4
Atomic Emission Spectra – Visible Electronic Transitions

Objectives
- Examine atomic emission spectra of several elements
- Understand how electron transitions in elements produce various forms of light

Topics Introduced in this Lab
- Atomic emission spectra
- Electron transitions in elements

Related Zumdahl sections
- Chapter 2 Sections 1, 2, 3, 4, 5, 6, and 7

Safety Considerations
- Metal ion solutions can be poisonous if ingested. Do not taste any of the metal ion solutions.
- The metal wire used to heat the solutions will be hot after it is inserted into the Bunsen burner. Do not touch it until it has cooled to room temperature.
- Be careful when inserting and removing elemental emission tubes from the DC power supply. Turn off DC power supply before removing one tube then turn it back on after inserting the next tube. DC power supplies operate at high voltage. Emission tubes will be hot after operation.
- Wear safety glasses and appropriate lab wear at all times while in the lab.

Introduction
Electrons occupy quantized energy states in atoms. Scientists designate those quantized energy states using electron configurations. For example, in its lowest energy (ground) state the one electron in an H atom is in the 1s quantum state ($1s^1$ electron configuration). Li atoms contain three ground state electrons in the 1s and 2s states, two electrons in the 1s and one in the 2s ($1s^2\ 2s^1$ electron configuration). In Na the eleven ground state electrons occupy 1s, 2s, 2p, and 3s states ($1s^2\ 2s^2\ 2p^6\ 3s^1$ electron configuration).

When appropriate forms of energy (heat and electricity being two commonly used energy sources) are applied to atoms, electrons transition to other higher, excited energy states. To make these transitions electrons must absorb energy. Once in the excited energy state, electrons can exist there briefly, a few nano- to microseconds, then return to the ground state by emitting quantized energy units. Electrons do not necessarily return to the ground state in a single transition but may instead do so in several smaller transitions whose energies sum to that of the absorbed energy. Energy emission is accomplished by the generation of light particles, photons, in the infrared, visible, and ultraviolet portions of the electromagnetic spectrum. In other words, electron transitions from excited states to the ground state can produce light that is visible to our eyes. Among many others, the makers of fireworks, compact fluorescent lights, and street lamps understand this, utilizing the effect to make their products.

Electrons cannot transition to every available higher or lower lying energy state because they must obey quantum mechanical laws. Of importance for this discussion, we must recognize that photons have one unit of angular momentum. When the electron transition occurs, the angular momentum (ℓ) of the start and end states must differ by at least one unit of angular momentum which is imparted by the photon. Therefore, the 2s to 1s transition is not allowed, but the 2p to 1s transition is. (The p orbital has angular momentum $\ell = 1$ while the s orbital has angular momentum $\ell = 0$.) Some (but not all) possible transitions are illustrated in **Figure 1**.

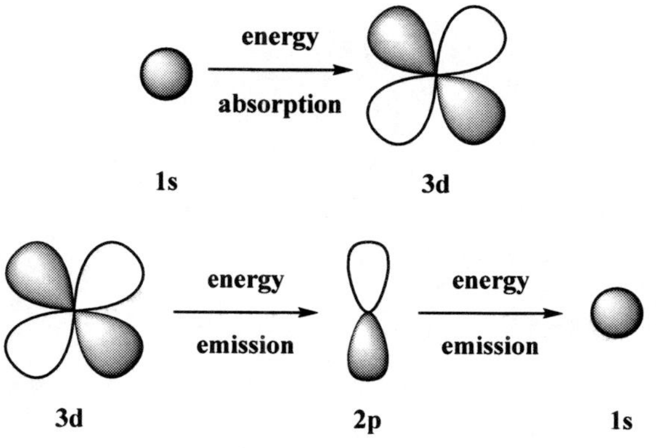

Figure 1. Possible Atomic Electron Absorption and Emission Transitions

$$E = h\nu = \frac{hc}{\lambda}$$

where E is the energy in J or kJ

h is Planck's constant $= 6.626\times10^{-34}$ J·s

c is the velocity of light $= 3.00\times10^{8}$ m/s

ν is the frequency of light in $1/\mathrm{s}$ (Hz)

λ is the wavelength of light in m

Equation 1. Planck Equation

When heated, Na produces yellow-colored emission photons having a wavelength of 589.0 nm. These photons originate from the transition of electrons from the 3p to the 3s level. Using **Equation 1** and the measured wavelength we can determine the energy of the emitted photons.

$$\lambda = 589.0\text{ nm}\left(\frac{1\times10^{-9}\text{m}}{1\text{ nm}}\right) = 5.89\times10^{-7}\text{m}$$

$$E = \frac{hc}{\lambda} = \frac{(6.6.26\times10^{-34}\text{J}\cdot\text{s})(3.00\times10^{8}\text{m/s})}{5.89\times10^{-7}\text{m}} = 3.37\times10^{-19}\text{J}$$

This 3.37×10^{-19} J, an incredibly small amount of energy, is the energy contained in a single photon of this yellow-colored light. Our eyes are insensitive to such a small quantity of energy. We simply do not have a vision system that can detect so little light. However, if 60 or so photons having this amount of energy arrive in our eyes at the same time, we can see that as small flash of yellow light. Let's calculate the amount of energy necessary for our eyes to perceive this yellow light.

$$60\text{ photons}\left(\frac{3.37\times10^{-19}\text{J}}{\text{photon}}\right) = 2.0\times10^{-17}\text{ J}$$

It's incredible to believe that our eyes are sensitive enough to light that we can actually see something containing so little energy.

Procedure

Part A – Seeing Atomic Emission Spectra

1. Pour a thin layer of 0.1 *M* LiCl(*aq*) into the bottom of a Petri dish and allow the water to evaporate. If necessary slightly heat the Petri dish to increase the evaporation. This will leave a residue of solid LiCl in the bottom of the Petri dish.
2. In a separate Petri dish repeat step 1 using 0.1 *M* NaCl.
3. In a third Petri dish repeat step 1 using 0.1 *M* KCl.
4. In a fourth Petri dish repeat step 1 using 0.1 *M* $CuCl_2$.
5. In a fifth Petri dish repeat step 1 using 0.1 *M* $SrCl_2$.
6. Either in a fume hood or an area of the lab away from flammable material, use a squirt bottle containing liquid methanol to place a thin layer of methanol into each of the Petri dishes, **Figure 2**. Remove the bottle of methanol from your work area before performing step 7.

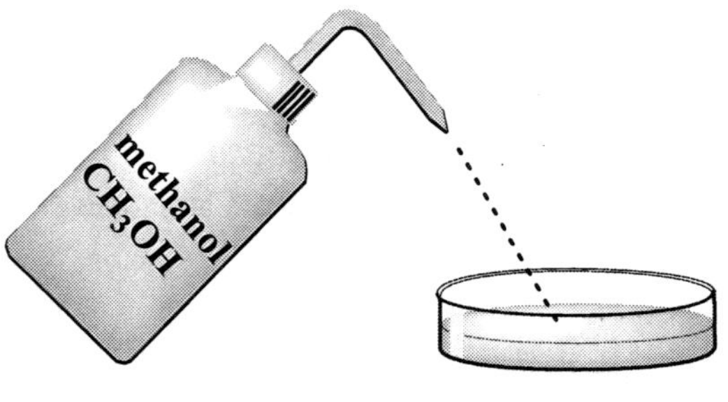

Figure 2. Placing Methanol into Petri Dish

7. Using a match or lighter, ignite the methanol in each of the Petri dishes. Use extra caution when igniting the methanol.
8. Methanol burns with a light blue-colored flame. Once the blue flame is replaced by a different color, the salt has heated to the point it is emitting its characteristic atomic spectrum. You may need to turn off the room lights to better see the colors.
9. Stand back from the Petri dishes and observe the colors emitted by each salt.
10. Record the flame colors of each salt.

Part B – Spectrometric Determination of Atomic Spectra

11. Partner with another lab student to perform the remainder of this experiment.
12. Obtain a spectrophotometer or diffractometer, Bunsen burner, and piece of Nichrome (or other suitable wire) from your instructor.
13. Form the Nichrome or other suitable wire into a long section with a loop on the end similar to this image.

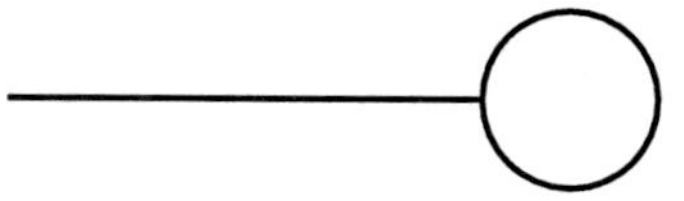

14. Pour 10 to 15 mL of the 0.1 *M* LiCl(aq) solution into a small beaker.
15. Repeat step 14 for each of the other salt solutions using separate labeled beakers for each solution.
16. Ignite the Bunsen burner, adjusting the air and gas mixture to produce a flame having an outer light blue cone surrounding a darker blue cone of flame.
17. One partner must submerge the wire loop into the 0.1 *M* LiCl(aq) solution, then withdraw the loop from the solution and place it into the Bunsen burner flame, **Figure 3**. As the emission spectrum is emitted the second partner should look at the spectrum using the spectrophotometer or diffractometer. Count the number of visible emission lines in the spectrum and record that number below.
18. Switch places with your partner so that both can see the spectrum.
19. Repeat steps 17 and 18 for each of these salt solutions, 0.1 *M* NaCl, 0.1 *M* KCl, 0.1 *M* $CuCl_2$, and 0.1 *M* $SrCl_2$.
20. If your lab room has fluorescent or compact fluorescent lights, look at them using your spectrophotometer or diffractometer.

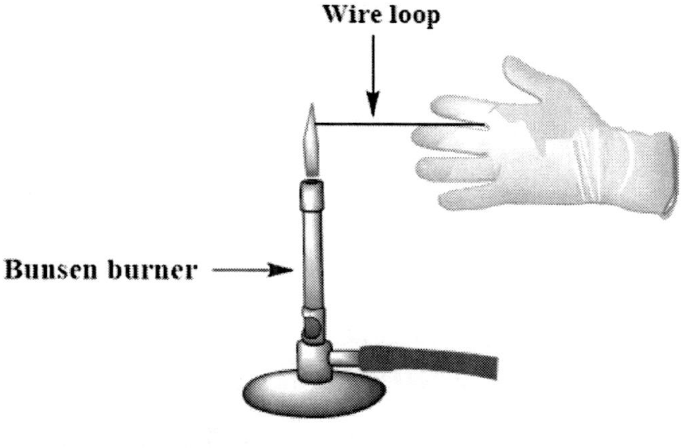

Figure 3. Holding Wire Loop in Bunsen Burner Flame

Part C – Measuring a Few Atomic Spectra

21. In the lab room you will find some of these elemental emission tubes (H, He, Hg, Ne, or Ar) which can be placed into a DC power supply. After one of the bulbs is inserted into the power supply and it is turned on the visible spectrum of that element is emitted. **Be careful to turn off the DC power supply before you remove and insert another elemental emission tube. These power supplies operate at high voltage! Tubes are hot after they have been in power supply for a few minutes.**
22. Using your standardized spectrophotometer, which can measure the wavelengths of each emission line, look in turn at each bulb and record the wavelengths of the emitted light from each element.
23. Compare the spectrum emitted by the fluorescent lights which you saw in Part B to the elemental spectra seen in Part C. If a Part C spectrum matches the fluorescent light spectrum can you identify which element is contained in the fluorescent light bulbs?

Experiment 4 Pre-Lab (Show ALL Work and ALWAYS include units)

1. Indicate which of these electron transitions are not quantum mechanically possible by placing a check mark in the appropriate blank.

	Possible	Not Possible
2s → 1s	____	____
3p → 2s	____	____
3p → 2p	____	____
3d → 2p	____	____
3p → 3s	____	____

2. For the sake of this problem, let's assume that 2.0×10^{-17} J is the smallest energy of visible light that humans can see. A typical green laser pointer has a wavelength of 532 nm. How many photons of this green light are necessary for us to see a brief green flash?

3. A typical green-colored laser pointer has a wavelength of 532 nm and a power of 0.500 mW. (mW is milliwatts, which is equivalent to 0.001 watts. 1.00 watt = 1.00 J/s.) If your instructor points this laser onto a projector screen for 2.5 seconds, how many green light photons will strike the projector screen in that time frame?

4. How does the existence of discrete atomic emission spectra indicate that quantum mechanics operates in atoms?

Experiment 4 Lab Report (Show ALL Work and ALWAYS include units)

1. Color of LiCl flame ________
2. Color of NaCl flame ________
3. Color of KCl flame ________
4. Color of $CuCl_2$ flame ________
5. Color of $SrCl_2$ flame ________
6. Number of visible lines in LiCl emission spectrum ________
7. Number of visible lines in NaCl emission spectrum ________
8. Number of visible lines in KCl emission spectrum ________
9. Number of visible lines in $CuCl_2$ emission spectrum ________
10. Number of visible lines in $SrCl_2$ emission spectrum ________
11. Record the wavelengths of the visible lines for each of the elements that you examined in the elemental emission tubes.

Element **Line Wavelengths**

H

He

Hg

Ne

Ar

12. In its emission spectrum Li has the following prominent lines. (There are also other lines that are omitted from this discussion.) The 460 nm line is generated by an electron making the transition from the $1s^24d$ energy state to the $1s^22p$ energy state. The other four lines are generated by these transitions:

$1s^25d$ to $1s^22p$
$1s^24s$ to $1s^22p$
$1s^22p$ to $1s^22s$
$1s^25s$ to $1s^22p$

Assign each transition to its correct emission line.

Wavelength (nm)	Color
671	red
497	green
460	green
427	blue
413	blue

Transition **Wavelength (nm)**

$1s^25d$ to $1s^22p$:

$1s^24s$ to $1s^22p$:

$1s^22p$ to $1s^22s$:

$1s^25s$ to $1s^22p$:

Experiment 5
Photonic Crystals –
Creation of a Synthetic Inverse Opal Photonic Crystal

Objectives

- See a photonic crystal
- Understand the science behind photonic crystals

Topics Introduced in this Lab

- Crystal sizes affect how light is reflected and diffracted
- Constructive and destructive interference
- Indices of refraction
- Periodic crystalline structures can cause light to behave in an understandable fashion

Related Zumdahl sections

- Chapter 2 Sections 1, 3, and 5

Safety Considerations

- Piranha solution is a 3:1 mixture of concentrated sulfuric acid and 30% hydrogen peroxide. If this solution comes into contact with your skin, immediately flush it with copious amounts of water. Be certain to wear safety glasses as piranha solution in your eyes could lead to permanent eye damage. Handle with care. Gloves are recommended.
- If any of these compounds come into contact with your eyes or skin, immediately flush the affected area with water.
- Do not ingest any of these compounds.
- Wear safety glasses and appropriate lab wear at all times while in the lab.

Introduction

Early in the history of the discovery of atomic structure, scientists realized that solid crystalline structures have the ability to diffract electromagnetic radiation. Remember, electromagnetic radiation is all of the various light forms including gamma rays, x-rays, ultraviolet, visible, infrared, and microwave radiation. William Henry Bragg and his son William Lawrence Bragg received the 1915 Nobel Prize in Physics for their understanding that x-ray diffraction in solid crystals was the result of the crystalline geometry providing constructive interference of waves that matched the crystal spacing. Constructive interference is also possible if the crystal spacing is an integral multiple (twice, three times, etc.) of the wavelength. Waves that do not meet this matching condition destructively interfere with one another, causing them to disappear, see **Figure 1**. The important point to understand is that waves with wavelengths that match the crystal spacing are enhanced while those that do not are eliminated.

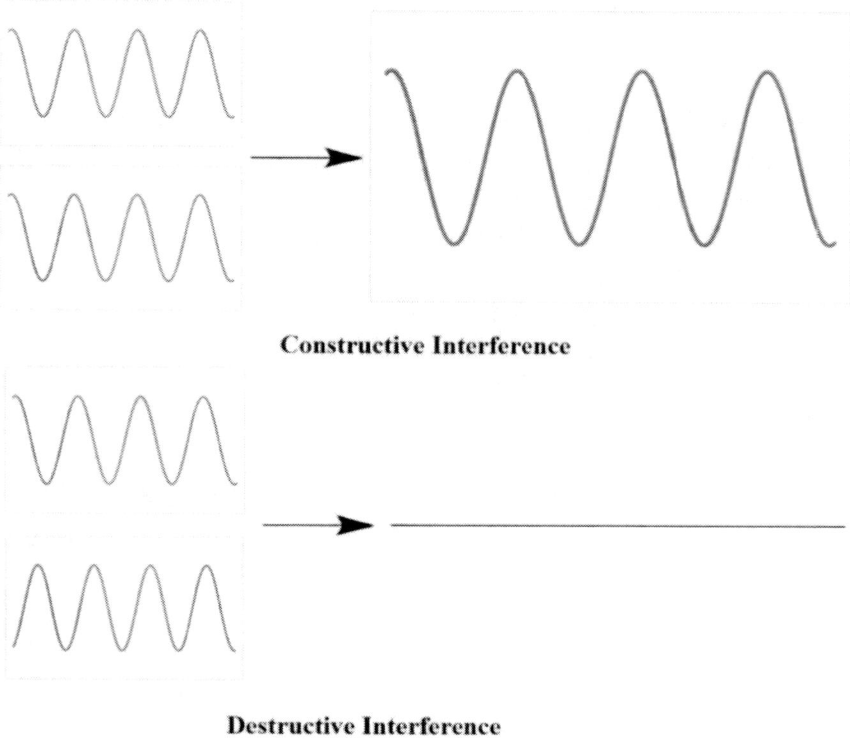

Figure 1. Constructive and Destructive Interference of Two Waves

Crystalline solids have a regular repeating pattern, called a unit cell, that can have a variety of shapes. Cubic, tetragonal, monoclinic, orthorhombic, rhombohedral, hexagonal, and triclinic shapes are all possible. As long as the unit cell produces a regular periodically repeating pattern, constructive wave interference will occur if the particle spacing matches the wavelength of the radiation (light) that is directed upon the solid. In **Figure 2** you can see as an example that a cubic unit cell produces such a repeating pattern.

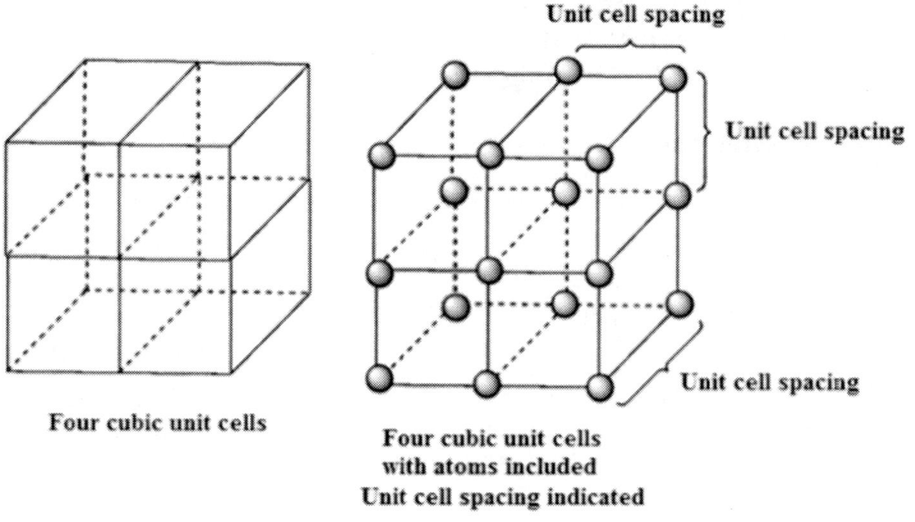

Figure 2. Four Cubic Unit Cells – Same Unit Cells with Atoms and Cell Spacing Indicated

X-rays have wavelengths of 10×10^{-12} m (10 picometers, pm) to 10×10^{-9} m (10 nanometers, nm). Metal atoms typically have radii of ~100 pm. For example, solid metallic nickel (Ni) has a resulting unit cell spacing of 325 pm (see **Figure 2**). Many solid crystal unit cells have spacings that lie in the range of x-ray wavelengths. Consequently, x-rays are ideal for probing crystal lattices because of the matching wavelength and cell spacing distances. Photographic or electronic images of the pattern that results from x-rays interacting with a crystal are used by scientists to determine crystal structures.

Building on this idea, by making regularly spaced patterns of various sizes, those patterns can constructively/destructively interfere with a variety of different wavelengths. For instance, visible light has wavelengths from 450 to 700 nm. This size range is bigger than individual atoms and molecules but much smaller than macroscopic objects. However, this range *is* in the realm of the relatively new field of nanoparticles. Theoretically, if we could make a repeating pattern of solid nanoparticles, it would produce a crystal, called a photonic crystal, where visible light waves constructively and destructively interfere, producing a brilliantly colored solid.

As so often occurs, nature has beaten us to the punch. October's birthstone, opal, is a photonic crystal consisting of silica spheres composed of SiO_2 molecules. A bright green-colored beetle found in Brazil, *Lamprocyphus augustus*, owes its color to photonic crystals formed in the beetle's scales from chitin molecules. As visible light passes through these two photonic crystals it encounters alternating patterns of solid (silica or chitin) and air. Light's passage from solid to air over and over in a regular pattern is the final piece of the photonic crystal puzzle.

As light travels from one medium (solid, liquid, or gas) to a different medium its velocity changes and it is refracted (bent). Physical measurement of this phenomenon results in a property called a refractive index, which is the ratio of the velocity of light in vacuum divided by the velocity of light in the medium, **Equation 1**.

$$\textit{refractive index} = c/v$$

where c = velocity of light in vacuum
v = velocity of light in another medium

Equation 1. Refractive Index Equation

For example, water has a refractive index of 1.33, indicating that in vacuum light has a velocity 1.33 times greater than in water. When sunlight encounters raindrops refraction occurs, resulting in a rainbow.

Photonic crystals result when light encounters a periodic pattern of different refractive indices. For example, if light passes from solid to air to solid to air in a regular pattern, as indicated in **Figure 3,** this could be an indication of a photonic crystal. The one remaining requirement is that the repeating pattern has a spacing that is one half the wavelength of the refracted light. Constructive wave interference occurs for the refracted light while other wavelengths destructively interfere and are eliminated from view, see **Figure 4**.

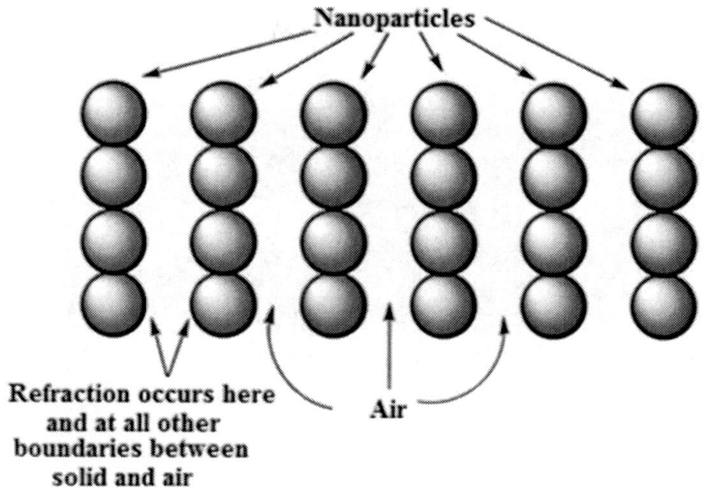

Figure 3. Representation a Photonic Crystal at the Nanoparticle Level

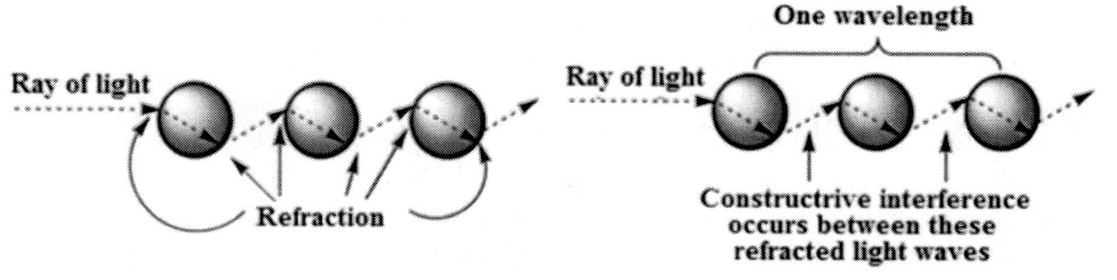

Figure 4. Representation of Refraction and Constructive Interference in a Photonic Crystal

In this experiment we will create a photonic crystal with a structure similar to that of opals, called an inverse opal. Rather than use silica molecules to make the nanoparticles we will use a polymeric form of styrene, **Figure 5**. Under appropriate conditions, spherical nanoparticles of polystyrene precipitate, forming a solid consisting of voids and solids that produce a photonic crystal, **Figure 6**.

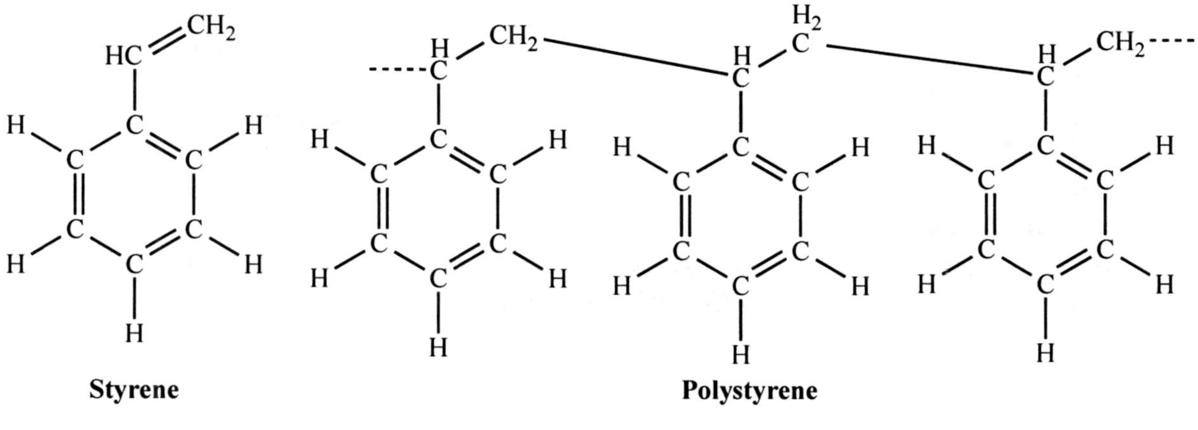

Figure 5. Chemical Structure of Styrene and Polystyrene

Figure 6. Scanning Electron Microscope Image of Inverse Opal Photonic Crystal Assembly

Procedure

Part A – Glass Slide Preparation

For the inverse opal to properly adhere to the microscope slide the glass slide must be cleaned in several solutions that make the glass highly hydrophilic (water loving).

1. Soak the glass microscope slide in acetone for 5 minutes, stirring the acetone the whole time. Carefully rinse the glass slide with deionized water then dry with a Kim-wipe or other suitable paper wipe.
2. Soak the slide for 5 minutes in isopropyl alcohol, stirring the alcohol the whole time. Rinse the slide again with deionized water and dry with a Kim-wipe or other suitable paper wipe.
3. Place the slide in the Piranha solution for approximately 30 minutes. Carefully rinse the slide with deionized water, rinse with isopropyl alcohol, then wipe dry with a Kim-wipe or other suitable paper wipe. **(Piranha solution is a 3:1 by volume mixture of concentrated sulfuric acid and 30% hydrogen peroxide. If you get any of this solution on your skin it will rapidly react with your skin. Treat this solution with great care. Wash any part of your skin that comes in contact with Piranha solution with copious amounts of water.)**
4. Dry the glass slide in an oven for 30 minutes.

Part B - Photonic Crystal Fabrication

5. Using a 10-mL syringe, minus the needle, place 10 mL of deionized water into a shell vial (small beaker with a top that is not rounded).
6. Using a micropipette, pipet 200 μL (200 microliters) of polystyrene solution into the shell vial. Using a stirring rod mix the solution until it is homogeneous.
7. Use tweezers to pick up the cleaned glass slide and place it vertically into the polystyrene/water mixture. (See picture below.)

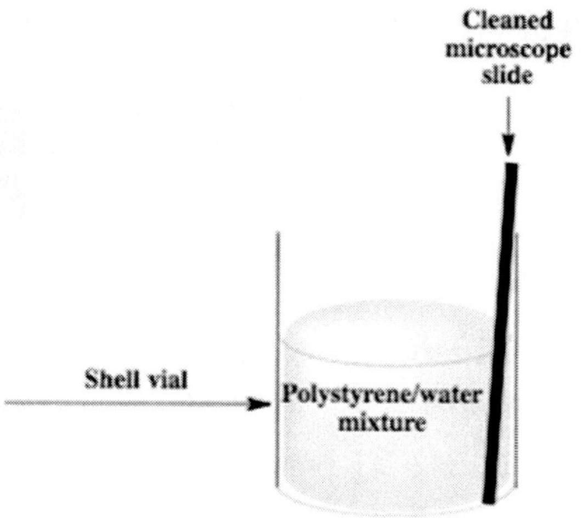

8. Place the shell vial into an oven for 3 hours, allowing the mixture to slowly and completely evaporate. Carefully remove the glass slide and allow it to return to room temperature.
9. Visually examine the slide and record the color(s) you see in your lab manual.

Experiment 5 Pre-Lab (Show ALL Work and ALWAYS include units)

1. If we wish to make a photonic crystal that produces yellow-colored light having a frequency of $5.26 \times 10^{14}\ s^{-1}$, what must be the distance (in nm) between the solid spheres in the photonic crystal?

2. Explain why constructive interference enhances a wave.

3. Explain why destructive interference destroys a wave.

4. Is it possible for a substance to have a refractive index that is less than one? Explain your answer.

Experiment 5 Lab Report (Show ALL Work and ALWAYS include units)

Photonic Crystal Preparation

1. Describe the color of the polystyrene/water mixture.

2. Describe the color of the glass slide when you remove it from the polystyrene/water mixture.

3. Describe the color of the glass slide once removed from the oven.

4. To produce the observed colors on the glass slide what must be the size of the particle spacing in the photonic crystal? Explain.

Experiment 6
Making Infrared Flares – A Self-Directed Experiment

In their tool inventory, US soldiers use nighttime vision goggles that allow them to see other soldiers at night. Nighttime vision goggles work because the goggles' lenses convert infrared light into visible light. Infrared light is usually emitted as heat. (For example, French fries at a fast food restaurant are frequently kept under an infrared lamp to maintain their warmth.) Any object having a temperature a few degrees different than nighttime ambient temperature is visible to a soldier wearing nighttime vision goggles. However, that still does not provide soldiers with sufficient clarity to prevent running into bushes, logs, etc. that are at the same temperature as nighttime air. It would be nice if soldiers could fire a flare into the sky that emits only infrared and no visible light. This would illuminate the field for US soldiers but not for others.

In this experiment, you will design and perform tests to determine which metal nitrate salt(s) (from a list of possible salts) emits primarily or exclusively infrared light and little or no visible light when burned in air. You must also decide which metal nitrate salt is most cost effective for making these flares. A list of metal nitrate salts with their costs is given in **Table 1**. You will have access to a few grams of each salt, an IR-visible spectrophotometer, and other necessary lab equipment.

Salt	Price (per kilogram)
Lithium nitrate	$190
Sodium nitrate	$75
Potassium nitrate	$100
Cesium nitrate	$936
Magnesium nitrate	$116
Strontium nitrate	$170
Barium nitrate	$246

Table 1. Metal Nitrate Salts and Their Costs

Prior to the experiment, you must write your proposed procedure and present it to your instructor. He or she will read your proposed procedure, make suggestions on the proposal while ensuring that the procedure is safe, and then return it to you prior to the experiment. You are expected to perform the procedure on your own without assistance from other students or the instructor. After you perform the experiment, you will write a lab report that describes the procedure you used, the results you obtained, and the conclusions you can draw from your results. Ask your instructor about any specific format guidelines for writing the proposed procedure and/or the lab report.

Experiment 7
Understanding and Using Absorption Spectroscopy

Objectives
- Understand that light absorption promotes excitation of electrons.
- Explore the relationship between absorbance and concentration of a solution (Beer-Lambert law).
- Use the Beer-Lambert law to determine the concentration of a solution.

Topics Introduced in this Lab
- Absorption of light by atoms or molecules
- Molarity as a unit of concentration
- Solutions

Related Zumdahl Sections
- Chapter 2 Sections 1 and 3
- Chapter 6 Section 3

Safety Considerations
- Safety goggles and appropriate lab attire must be worn at all times when working in the lab.
- Don't ingest any of the chemicals in this experiment.
- Copper sulfate solutions are toxic and should be handled with care. If you get any on you, rinse immediately with copious amounts of water and then wash thoroughly with soap and water.

Introduction
Recall that light is emitted at specific wavelengths when electrons transition from a higher energy level in a chemical species to a lower energy level. The wavelength of that light corresponds to an amount of energy equal to the difference in energy between the two levels occupied by the electron. In a similar way, light can promote electrons from a lower energy level to a higher one, if the wavelength of the light corresponds to the change in energy between those two levels. This is how light is absorbed by a chemical species.

Solutions of chemical compounds can be made by dissolving the species of interest (the solute) in another chemical compound. This second compound, frequently a liquid, is called the solvent. Solutions are used widely in chemistry because they allow chemical species to easily approach one another so that they can undergo a chemical reaction. To successfully use solutions to run chemical reactions, it is essential to quantify the amount of a chemical species in a given volume of solution. Therefore, a unit of concentration is needed. The most common unit of concentration used in chemistry is molarity (M), which is defined as the moles of solute per liter of solution, **Equation 1**. It is important to recognize that this is liters of solution and NOT liters of solvent.

$$M = \frac{\textbf{moles of solute}}{\textbf{L of solution}}$$

Equation 1. Molarity

How much light is absorbed by a solution is dependent on the concentration of that solution. This is a fact that you probably have used even though you might never have realized it. Think about a solution made using water as the solvent and Kool-aid powder as the solute. If you were shown two bottles of Kool-aid made in this way, could you tell if one was more concentrated than the other? You would probably examine the two solutions and rightly say that the one appearing darker in color was more concentrated. The more concentrated solution has more absorbing species per unit volume and therefore absorbs more light, allowing less light to pass through the solution and the solution appears darker.

Intuitively it makes sense that more concentrated solutions will absorb more light (which is the same as saying that the solutions allow less light to pass through). Mathematically, this is expressed in the Beer-Lambert law equation, **Equation 2**. In this expression, A is the absorbance of the solution, ε is the molar absorptivity (a variable unique to a given species that takes into account that inherent differences in chemical compounds cause them to absorb different amounts of light), b is the distance light travels through the solution, and c is the concentration of the solution.

$$A = \varepsilon bc$$

Equation 2. Beer's Law

This proportional relationship between concentration and absorbance can be used to the advantage of the shrewd chemist. If multiple solutions are examined in which the path length of the light (b), and the identity of the absorbing species is kept constant (ε remains the same), then the relationship between absorbance and concentration simplifies even further to **Equation 3**. This expression resembles the form of a line, $y = mx + b$, where $A = y$ and $c = x$. Therefore, if the absorbance and concentration are plotted for a set of solutions, keeping b and ε constant, a linear relationship should result. Once this plot, known as a Beer-Lambert plot, has been created, the absorbance of a solution for which the concentration is unknown can be "plugged in" as the y value on the Beer-Lambert plot, to find the corresponding x value, which is the concentration!

$$A = (\textbf{constant}) \cdot c$$

Equation 3. Simplified Beer's Law

All of this is of course contingent on being able to measure the absorbance of solutions. To do this, we use an instrument called a spectrophotometer, which shines a beam of light through a sample of solution, **Figure 1**. A detector measures the amount of light that passes through the sample. If light is absorbed, less light will reach the detector than left the light source. From this percentage of light that reaches the detector, the amount of light absorbed is determined.

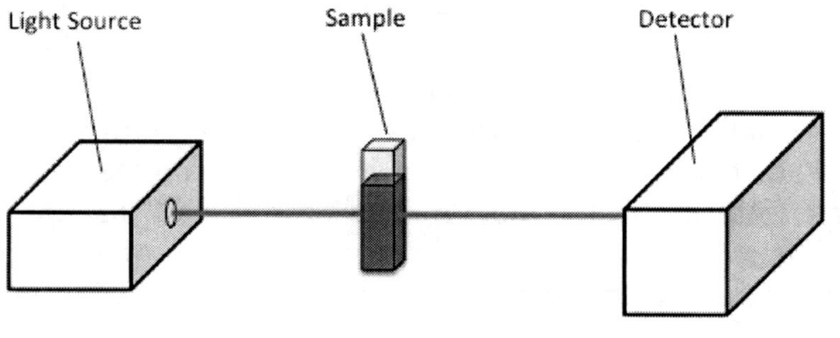

Figure 1. Basic Schematic of a Spectrophotometer

The spectrophotometer will measure and record the absorbance value at each wavelength of a designated set (such as all wavelengths in the visible range). This measurement produces an absorbance spectrum with absorbance versus wavelength data, **Figure 2**. If the solution absorbs light in the wavelength region specified, absorbance bands (broad peaks) will be seen. The lambda max (λ_{max}) is a wavelength at which maximum absorbance occurs and is therefore located at the wavelength corresponding to the peak of one of the absorbance bands. A λ_{max} is identified and absorbance values are then always recorded at that specific wavelength for a given experiment. *For absorbance values to be comparable, they must all be measured at the same* λ_{max} *value.*

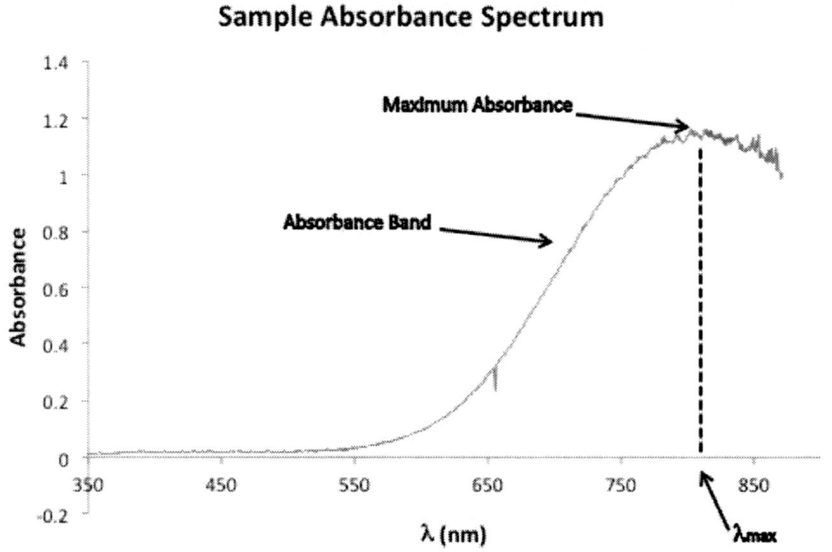

Figure 2. Sample Absorbance Spectrum

To create a Beer-Lambert plot, you first need a set of solutions where the concentrations of the absorbing species are known accurately. These concentrations should span a range that includes all possible concentrations of any unknowns you'll be examining. Usually, you need to make these "standard" solutions yourself. In this experiment however, the solutions are pre-made for you. There are five solutions that you will use to create a Beer-Lambert plot. The solutions have concentrations of 0.010 *M*, 0.025 *M*, 0.050 *M*, 0.075 *M*, and 0.10 *M*.

One last point must be considered when preparing to use absorption spectroscopy. Namely, the container used to hold the solution when absorption measurements are taken plays a key role in the experiment. First, this container, called a *cuvette*, determines the distance the light travels through the solution. In addition, the container changes the direction of some of the light that passes through it. Light can diffract, refract, and reflect when it encounters the container. This change in the direction of the light prevents it from reaching the detector and the misdirected light is therefore incorrectly attributed to absorbance by the solution sample, **Figure 3**. To negate this error (and before any absorbance measurements of the solution sample are taken) the cuvette is filled with only solvent and its "absorbance" is measured. This is called the reference cuvette or a blank cuvette. Light that does not reach the detector because it is diffracted, refracted, or reflected by the cuvette and solvent gives a false absorbance measurement. Since the same cuvette and solvent will be used for measurements of the solution samples, this false absorbance value can be subtracted from the absorbance measurements of samples, thereby subtracting out the "absorbance" due to path changes in the light and giving an absorbance reading that accurately reflects the absorbance of the species in solution. The absorbance of the reference can be measured prior to the absorbance of the solution sample or at the same time, depending on whether a single beam or dual beam spectrophotometer is used. A schematic of a dual beam spectrophotometer is shown in **Figure 4**.

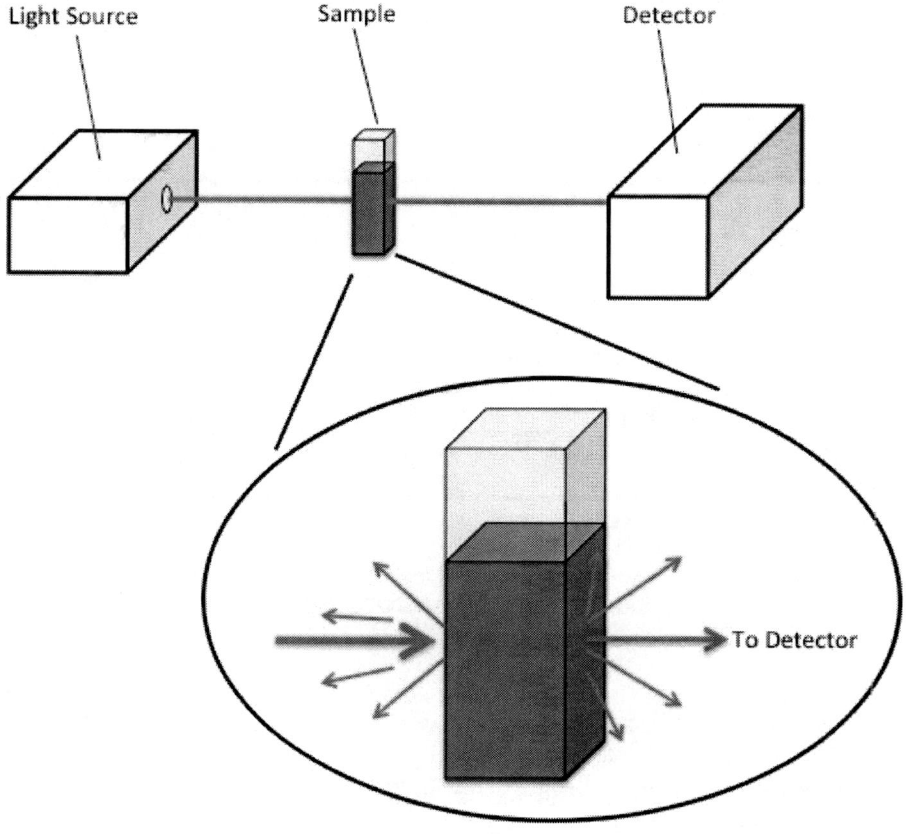

Figure 3. Effects of Diffraction, Refraction, and Reflection

To summarize, you will record absorbance values at the same λ_{max} value for each of the five standard solutions of $CuSO_4(aq)$ provided. You will plot these five absorbance values versus the concentration of the solutions to create a Beer-Lambert plot. You will then record an absorbance value (from the same λ_{max}) for a solution that you make, and using the Beer-Lambert plot, determine the concentration of that solution.

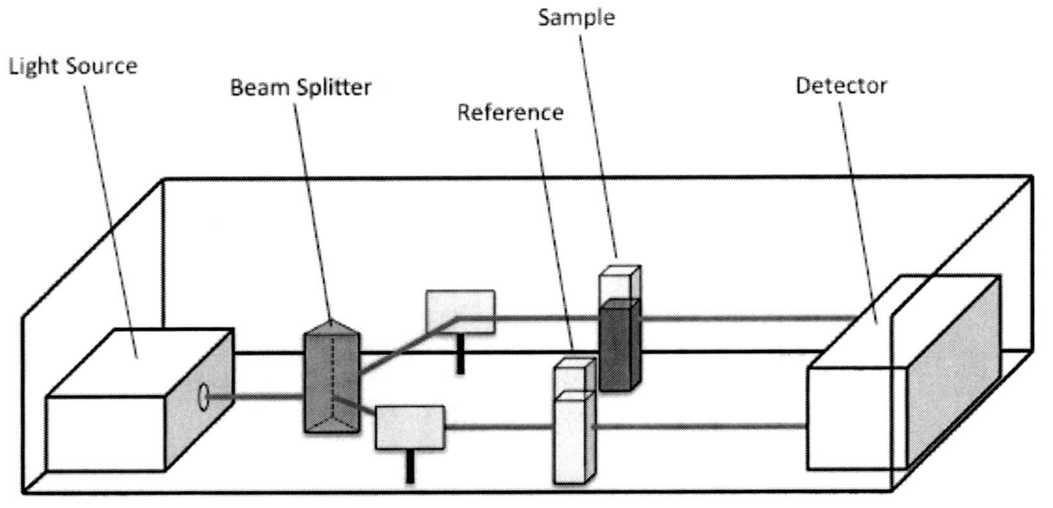

Figure 4. Schematic of a Dual Beam Spectrophotometer

Procedure

Part A – Measuring Absorbance Values of the Standard Solutions

1. Fill a cuvette with the 0.10 *M* $CuSO_4(aq)$ solution and then insert it into the spectrophotometer. Following your instructor's directions on using the spectrophotometer, record a spectrum of the solution from 450 nm to 850 nm. From this spectrum determine the λ_{max} for $CuSO_4(aq)$.
2. Set the spectrophotometer to the λ_{max} wavelength, then record the absorbance value for the 0.10 *M* $CuSO_4(aq)$ solution. Measure all subsequent solution absorbance values at this wavelength.
3. Dispose of the contents of the cuvette in the appropriate waste container and rinse it thoroughly. All rinse waste should also go into the waste container. The final rinse should be done with distilled or deionized water.
4. Using the same cuvette, repeat steps 2 and 3 for the remaining four solutions of known concentration. Be sure to record the measured absorbance value and concentration of each solution in your lab report.

Part B – Making and Recording Absorbance of Solution of Unknown Concentration

5. Carefully weigh out about 0.1 grams of $CuSO_4{\cdot}5H_2O$ into a 50-mL beaker. Record the exact mass used in your lab report. Add 10.0 mL of deionized or distilled water and stir with a stirring rod.
6. After all of the solid has dissolved, transfer some of your freshly prepared solution to your cuvette and record the absorbance value just as you did for your standard solutions. Be sure the absorbance is recorded from the same λ_{max} you used earlier.

Experiment 7 Pre-Lab (Show ALL Work and ALWAYS include units)

1. It is important to take absorbance readings at the same λ_{max}.

   (Circle One) True False

   Why?

2. Using the same cuvette throughout, you measure the absorbance of a set of standard solutions (all containing the same solute) and obtain the following results:

Concentration of Solution (*M*)	Absorbance @ λ_{max} (no units)
1.00	0.60
0.80	0.44
0.60	0.34
0.40	0.21
0.20	0.10

   In a graphing program like Excel©, create a Beer-Lambert plot by graphing the data in the table above. Absorbance values should be on the *y*-axis and concentration values on the *x*-axis in the scatter plot. After graphing the 5 data points and properly labeling your plot with a title, axis titles, and appropriate units, perform a linear fit analysis. Be sure to display the R^2 value and the equation for the linear fit on your plot. This Beer-Lambert plot should be printed out and turned in with your Pre-Lab questions.

3. Record the linear-fit equation from your Beer-Lambert plot in the space below.

4. Using the same cuvette that was used in question #2, you measure the absorbance for a solution of unknown concentration containing the same solute as the standard solutions above. The measured absorbance value is 0.28. Using this value and the linear-fit equation from the Beer-Lambert plot, determine the concentration of the solution. Show your work and answer in the space below. (Should you plug in the absorbance for *y* or *x* in your linear-fit equation?)

Experiment 7 Lab Report (Show ALL Work and ALWAYS include units)

Part A – Measuring Absorbance Values of the Standard Solutions

1. Value of chosen λ_{max}: ________ nm

2. Complete the following table for your standard solutions.

Concentration of Solution (*M*)	Absorbance @ λ_{max} (no units)

3. In a graphing program like Excel©, create a Beer-Lambert plot by graphing the data in the table above. Absorbance values should be on the *y*-axis and concentration values on the *x*-axis in the scatter plot. After graphing the 5 data points and properly labeling your plot with a title, axis titles, and appropriate units, perform a linear fit analysis. Be sure to display the R^2 value and the equation for the linear fit on your plot. This Beer-Lambert plot should be printed out and turned in with your lab report.

4. Record the linear-fit equation from your Beer-Lambert plot in the space below.

Part B – Making and Recording Absorbance of Solution of Unknown Concentration

5. Exact mass of $CuSO_4 \cdot 5H_2O$ used: ________ g

6. Absorbance of solution @ λ_{max}: ________

7. Using the measured absorbance and the linear-fit equation from your Beer-Lambert plot, determine the concentration of the solution you made. Show your work and answer in the space below. (Should you plug in the absorbance for *y* or *x* in your linear-fit equation?)

Experiment 8
Using Lewis Dot Structures and Models to Explore Molecular Structure

Objectives
- Practice drawing Lewis dot structures.
- Understand how Lewis structures and VSEPR theory help us predict molecular structure.

Topics Introduced in this Lab
- Lewis structures
- VSEPR theory

Related Zumdahl Sections
- Chapter 3 Sections 9, 10, and 11
- Chapter 4 Section 1

Safety Considerations
- Safety goggles and appropriate lab attire must be worn at all times when working in the lab.
- Do not ingest any model components. They pose a choking hazard.

Introduction
As chemists (and as chemistry students), the reason we are so interested in the electronic structure of atoms is because understanding electronic structure allows us to understand the bonding between atoms, and therefore predict molecular structure. An understanding of molecular structure, in turn, helps us explain and predict chemical properties.

In this lab, you will explore the bridge between knowing the electronic structure of atoms and deriving the structure of molecules. To do this, you will be "working backwards" using model kits to make molecular structures, which you will then use as guides to create Lewis dot structures. Lewis dot structures are simplistic representations of how electrons are shared (or not shared) between atoms in molecules or polyatomic ions. They use element symbols as atomic centers and represent shared (bonding) electrons with a line (–) and unshared (lone pair) electrons with two dots (:).

Simple Lewis dot structures can be used to predict three-dimensional molecular structure using Valence Shell Electron Pair Repulsion (VSEPR) theory. In its simplest form, VSEPR theory states that electron pairs, shared or unshared, "prefer" to be as far away from other pairs as possible. Intuitively, it makes sense that two centers of negative charge would repel each other, and that, therefore, the lowest energy arrangement of a molecule would be the one that maximizes the distance between electron pairs, both shared and unshared. After making your models and drawing your Lewis structures, you will be asked to use VSEPR theory to explain why the models have a given geometry.

Procedure (Both the procedure and lab report are designed to be completed in the lab)

Ball Color	Atom Represented
Black	Carbon
Red	Oxygen
Green	Chlorine
Blue	Nitrogen
White	Hydrogen

Table 1. Model Kit Reference Table

Part A - CCl_4

1. Obtain a model kit from your instructor. (Be sure to keep track of all model pieces and return the complete kit to your instructor at the end of the lab). In your model kit, balls of different colors represent atoms of different elements (see Table 1). The connectors each represent two shared electrons.
2. Choose the correctly colored balls to represent 4 chlorine atoms and 1 carbon atom. Connect them to create a model of carbon tetrachloride (CCl_4).
3. Draw a two-dimensional sketch of the model in the lab report in the designated area.
4. Answer the questions about CCl_4 in your report.

Part B - H_2O

5. Choose the correctly colored balls to represent 2 hydrogen atoms and 1 oxygen atom. Connect them to create a model of water (H_2O).
6. Draw a two-dimensional sketch of the model in the lab report in the designated area.
7. Answer the questions about H_2O in your report.

Part C - NH_3

8. Choose the correctly colored balls to represent 3 hydrogen atoms and 1 nitrogen atom. Connect them to create a model of ammonia (NH_3).
9. Draw a two-dimensional sketch of the model in the lab report in the designated area.
10. Answer the questions about NH_3 in your report.

Part D - CO_2

11. Choose the correctly colored balls to represent 2 oxygen atoms and 1 carbon atom. Connect them to create a model of carbon dioxide (CO_2). Keep in mind that all holes on each atom should be connected to some other atom.
12. Draw a two-dimensional sketch of the model in the lab report in the designated area.
13. Answer the questions about CO_2 in your report.

Part E - $COCl_2$

14. Without making a model, try to draw a correct Lewis structure for phosgene ($COCl_2$) in the designated area of the lab report.
15. Answer the questions in the lab report about the phosgene molecule before proceeding to step 16.
16. After drawing your Lewis structure and answering the related questions, choose the correctly colored balls to represent 2 chlorine atoms, 1 oxygen atom, and 1 carbon atom. Connect them to create a model of phosgene ($COCl_2$). Keep in mind that all holes on each atom should be connected to some other atom.
17. Answer the remaining questions in Part E of the lab report.

Part F - Additional Practice

18. Using your model kit as an aid where needed, draw correct Lewis structures in your lab report for the following compounds:
    a. Carbon monoxide (CO)
    b. Dichloromethane (CH_2Cl_2)
    c. Hydrazine (N_2H_4)
    d. Hydrogen peroxide (H_2O_2)
    e. Ethylene (C_2H_4)
    f. Ethyne (C_2H_2)

Experiment 8 Pre-Lab (Show ALL Work and ALWAYS include units)

1. How many valence electrons does a carbon atom have? ________

2. How many valence electrons does an oxygen atom have? ________

3. How many valence electrons does a nitrogen atom have? ________

4. How many valence electrons does a chlorine atom have? ________

5. What does VSEPR stand for?

6. What is the fundamental idea of VSEPR theory? (State this in your own words)

7. What does drawing Lewis structures and applying VSEPR theory to those structures allow chemists to do?

Experiment 8 Lab Report (Show ALL Work and ALWAYS include units)

Part A - CCl_4

1. Draw a two-dimensional sketch of your CCl_4 model below (with atoms as widely separated as possible) and then answer the questions that follow. Show your structure to your instructor before continuing.

2. Which atom in the molecule is least electronegative? ________

3. How many valence electrons does a Cl atom have? ________

4. How many valence electrons does a C atom have? ________

5. How many total valence electrons should there be since there are 4 Cl atoms and 1 C atom? ________

6. Each connector in your model represents two electrons being shared between atoms. How many shared electrons are there total in the CCl_4 molecule? ________

7. All shared and unshared pairs of electrons are valence electrons originally brought to the molecule by the individual atoms. Therefore, the number of shared and unshared electrons in the molecule must sum to the total number of valence electrons from the individual atoms. Keeping this in mind, how many unshared electrons must be present in CCl_4? ________

8. How many ***pairs*** of unshared electrons must be present in a molecule of CCl_4? ________

9. Atoms in a molecule *normally* bond in such a way that they achieve a filled valence shell of electrons. This is called the octet rule, since a filled valence shell usually consists of 8 electrons. ALL shared and unshared electrons associated with an atom can contribute to a filled valence shell.* How many shared electrons are around C in your sketch? ________
   *All electrons in a bond to or from an atom are treated as belonging to that atom. This means that all electrons shared between atoms are counted as belonging to both atoms.

10. Based on your answer to the above question, does C have a filled valence shell (does it obey the octet rule)? ________

11. How many shared electrons does each Cl atom have? ________

12. Does each Cl atom have a filled valence shell? ________

13. If not, how many more electrons does each Cl atom need to obey the octet rule? ________

14. Based on your answers to questions #7 and #13, are there enough unshared electrons left to satisfy the octet rule for each Cl atom? ________

15. Keeping the same shape, re-draw your sketch from question #1 using the element symbols in place of the atoms. Draw lines between atoms (symbols) to indicate shared electrons. Distribute unshared electrons around atoms (symbols) in pairs using two dots per pair. Distribute these so that the octet rule is satisfied for all atoms. Show your structure to your instructor before continuing.

16. Congratulations! You just drew your first Lewis dot structure. Based on what you have done, complete the following (preliminary) rules for writing Lewis structures.

The number of shared and unshared electrons in a Lewis structure must equal the number of __________ electrons from every __________ in the molecule.

The central atom is usually the __________ electronegative atom. Atoms (represented by element symbols) should be distributed around the central atom symmetrically and lines should be used to connect the central atom to the other atoms. Each line represents a __________ pair of electrons.

The remaining electrons should be distributed around atoms in pairs (represented by two dots) so that the __________ rule is fulfilled for every atom in the molecule. Two dots represent an __________ pair of electrons.

17. CCl_4 has a **tetrahedral** molecular geometry because it has 4 atoms evenly spaced (109.5 degrees apart) around the central atom. Using VSEPR theory, explain why carbon tetrachloride has this geometry.

Part B – H_2O

18. Draw a two-dimensional sketch of your H_2O model below and then answer the questions that follow. Show your structure to your instructor before continuing.

19. Which atom in the molecule is least electronegative? ________

20. How many valence electrons does a H atom have? ________

21. How many valence electrons does an O atom have? ________

22. How many total valence electrons should there be since there are 2 H atoms and 1 O atom? ________

23. Each connector in your model represents two electrons being shared between atoms. How many shared electrons are there total in the H_2O molecule? ________

24. All shared and unshared pairs of electrons are the valence electrons originally brought to the molecule by the individual atoms. Therefore, the number of shared and unshared electrons in the molecule must sum to the total number of valence electrons from the individual atoms. Keeping this in mind, how many unshared electrons must be present in H_2O? ________

25. How many *pairs* of unshared electrons must be present in a molecule of H_2O? ________

26. Atoms in a molecule *normally* bond in such a way that they achieve a filled valence shell of electrons. This is called the octet rule since a filled valence shell usually consists of 8 electrons. However, hydrogen achieves a filled valence shell with only 2 electrons. This is called the duet rule. ALL shared and unshared electrons associated with an atom can contribute to a filled valence shell.* How many shared electrons does each H have in your sketch? ________
*All electrons in a bond to or from an atom are treated as belonging to that atom. This means that all electrons shared between atoms are counted as belonging to both atoms.

27. Based on your answer to the above question, does H have a filled valence shell (does it obey the duet rule)? ________

28. How many shared electrons does the O atom have? ________

29. Does the O atom have a filled valence shell (does it obey the octet rule)? ________

30. If not, how many more electrons does the O atom need to obey the octet rule? ________

31. Based on your answers to questions #24 and #30, are there enough unshared electrons left to satisfy the octet rule for the O atom? ________

32. Keeping the same shape, re-draw your sketch from question #18 using the element symbols in place of the atoms. Draw lines between atoms (symbols) to indicate shared electrons. Distribute unshared electrons around atoms (symbols) in pairs using two dots per pair. Distribute these so that the octet rule is satisfied for O and the duet rule is satisfied for H. Show your structure to your instructor before continuing.

33. Based on what we have observed while drawing the Lewis structure of H_2O, we need to modify some rules for writing Lewis structures. Complete the following modifications to our original rules.

The central atom is usually the __________ electronegative atom. However, __________ atoms are never the central atom. Any unshared electrons in a structure should be distributed around atoms in pairs (represented by two dots) so that the ____________ rule is fulfilled for every atom in the molecule, except H, which achieves a filled valence shell with only __________ electrons. This is called the __________ rule.

34. Since H achieves a filled valence shell with fewer electrons than other elements, explain the first rule in Question #33 above.

35. H_2O has a **bent** molecular geometry because it has 2 atoms (104.5 degrees apart) around the central atom. Using VSEPR theory, explain why water has this geometry instead of a **linear** geometry (where the 2 outside atoms are 180 degrees apart around the central atom). (Hint: Consider the lone pairs of electrons and review the VSEPR theory paragraph in the introduction.)

36. Just like CCl_4, H_2O has four things around the central atom. However, the identity of those things is different. While CCl_4 has four atoms around the central atom, H_2O has 2 atoms and 2 lone pairs of electrons. Note that in the bent geometry, the angle between the atoms is less than the angle between atoms in the tetrahedral geometry. Based on this information, what can you deduce about the space taken up by a lone pair of electrons versus a bonding or shared pair of electrons in the VSEPR model?

Part C – NH_3

37. Draw a two-dimensional sketch of your NH_3 model below and then answer the questions that follow. Show your structure to your instructor before continuing.

38. How many valence electrons does a H atom have? _______

39. How many valence electrons does a N atom have? _______

40. How many total valence electrons should there be since there are 3 H atoms and 1 N atom? _______

41. Each connector in your model represents two electrons being shared between atoms. How many shared electrons are there total in the NH_3 molecule? _______

42. How many unshared electrons must be present in NH_3? _______

43. How many *pairs* of unshared electrons must be present in a molecule of NH_3? _______

44. How many shared electrons does each H have in your sketch? _______

45. Based on your answer to the above question, does H have a filled valence shell (does it obey the duet rule)? ________

46. How many shared electrons does the N atom have? ________

47. Does the N atom have a filled valence shell (does it obey the octet rule)? ________

48. If not, how many more electrons does the N atom need to obey the octet rule? ________

49. Are there enough unshared electrons left to satisfy the octet rule for the N atom? ________

50. Keeping the same shape, re-draw your sketch from question #37 using the element symbols in place of the atoms. Draw lines between atoms (symbols) to indicate shared electrons. Distribute unshared electrons around atoms (symbols) in pairs using two dots per pair. Distribute these so that the octet rule is satisfied for N and the duet rule is satisfied for H. Show your structure to your instructor before continuing.

51. NH_3 has a **trigonal pyramidal** molecular geometry because it has 3 atoms (107 degrees apart) around the central atom. Using VSEPR theory, explain why ammonia has this geometry instead of a **trigonal planar** geometry (where the 3 outside atoms are 120 degrees apart and in the same plane around the central atom).

52. Just like CCl_4 and H_2O, NH_3 has four things around the central atom. However, the identity of those things is different. While CCl_4 has four atoms around the central atom and H_2O has 2 atoms and 2 lone pairs of electrons, NH_3 has 3 atoms and 1 lone pair of electrons. Note that in the trigonal pyramidal geometry, the angle between the atoms is less than the angle between atoms in the tetrahedral geometry but greater than the angle between atoms in the bent geometry. Explain this "intermediate" bond angle using the VSEPR model.

Part D - CO_2

53. Draw a two-dimensional sketch of your CO_2 model below and then answer the questions that follow. Show your structure to your instructor before continuing.

54. How many valence electrons does a C atom have? _______

55. How many valence electrons does an O atom have? _______

56. How many total valence electrons should there be since there are 2 O atoms and 1 C atom? _______

57. Each *connector* in your model is composed of two electrons being shared between atoms. How many shared electrons are there total in the CO_2 molecule? _______

58. How many unshared electrons must be present in CO_2? _______

59. How many *pairs* of unshared electrons must be present in a molecule of CO_2? _______

60. How many shared electrons does the C have in your sketch? _______

61. Based on your answer to the above question, does C have a filled valence shell (does it obey the octet rule)? ________

62. How many shared electrons do each of the O atoms have? ________

63. Do the O atoms have a filled valence shell (does each obey the octet rule)? ________

64. If not, how many more electrons does each O atom need to obey the octet rule? ________

65. Are there enough unshared electrons left to satisfy the octet rule for each O atom? ________

66. Keeping the same shape, re-draw your sketch from question #53 using the element symbols in place of the atoms. Draw lines between atoms (symbols) to indicate shared electrons (being sure to use a separate line for each shared pair). Distribute unshared electrons around atoms (symbols) in pairs using two dots per pair. Distribute these so that the octet rule is satisfied for all atoms. Show your structure to your instructor before continuing.

67. CO_2 has a **linear** molecular geometry because it has 2 atoms 180 degrees apart around the central atom. Using VSEPR theory, explain why carbon dioxide has this geometry instead of a **bent** geometry like water.

68. The structure of CO_2 has 4 shared electrons between each O and C atom. These *double bonds* must sometimes be drawn in Lewis structures to complete the octet rule for all atoms using the limited supply of electrons available. To clearly illustrate this, try drawing a Lewis structure for CO_2 below ONLY USING SINGLE BONDS that fulfills the octet rule for all atoms. Show at least 2 attempted structures.

69. Based on what we have learned from the structure of CO_2 let's add an additional rule for drawing Lewis structures. Complete the following:

After connecting atoms with lines representing single bonds and distributing remaining electrons as unshared pairs, if the __________ rule cannot be completed for some/all atoms, lone pairs of electrons can be changed to shared pairs of electrons between atoms in a way that completes the octet for all atoms. Sharing 2 pairs of electrons between two atoms will create __________ bonds, and sharing 3 pairs of electrons between two atoms will create __________ bonds.

Part E - $COCl_2$

70. Without making a model, try to draw a correct Lewis structure for phosgene ($COCl_2$). Remember to determine which atom is central and determine the total number of valence electrons you have to distribute in the molecule.

71. Answer the following questions to determine if the structure you drew is a correct Lewis structure.

- What is the total number of valence electrons from 1 C, 1 O, and 2 Cl atoms? ________
- How many electrons (shared and unshared) did you use in your Lewis structure? ________
- If the above two numbers do not match, redraw your structure and answer the above questions again.
- Do each of the atoms in the structure have a filled octet? ________
- If not, redraw your structure and answer ALL of the above questions again.

72. Using your Lewis structure and the information found in question #51, determine whether phosgene has a **trigonal pyramidal** or **trigonal planar** geometry.

73. Using VSEPR theory, explain why phosgene has this geometry.

74. Make a model of phosgene according to step 16 of the *procedure*. Draw a two-dimensional sketch of your $COCl_2$ model below and then answer the questions that follow. Show your structure to your instructor before continuing.

75. Does your model confirm the geometry that was indicated by your Lewis structure and VSEPR theory?

76. Based on what you have learned in this lab (and especially your answer to the last question), why do we spend so much time learning how to draw Lewis structures and learning about VSEPR theory?

Part F – Additional Practice

77. Lewis structure for carbon monoxide (CO)

78. Lewis structure for dichloromethane (CH_2Cl_2)

79. Lewis structure for hydrazine (N_2H_4)

80. Lewis structure for hydrogen peroxide (H_2O_2)

81. Lewis structure for ethene (C_2H_4)

82. Lewis structure for ethyne (C_2H_2)

Experiment 9
Polar and Nonpolar Molecules – A Visual Experiment

Objectives

- See how polar molecules differ from nonpolar molecules.
- Experience the effect of molecular polarity on solubility.

Topics Introduced in this Lab

- Molecular polarity
- Solubility

Related Zumdahl sections

- Chapter 4 Sections 1 and 2
- Chapter 10 Section 1

Safety Considerations

- CS_2, carbon disulfide, has a noxious odor. It is also a skin and eye irritant.
- CH_2Cl_2, dichloromethane, is a skin and eye irritant.
- C_2H_5OH, ethanol, is highly flammable, and a skin and eye irritant. It is also an intoxicant. Do not expose this compound to an open flame.
- CH_3COCH_3, 2-propanone or acetone, is a skin and eye irritant. It is also highly flammable. Do not expose this compound to an open flame.
- C_6H_{14}, hexane, is highly flammable and a skin and eye irritant. Do not expose this compound to an open flame.
- $C_6H_4Cl_2$, ortho-dichlorobenzene, is a skin and eye irritant.
- If any of these compounds come into contact with your eyes or skin, immediately flush the affected area with water.
- Do not ingest any of these compounds.
- Wear safety glasses and appropriate lab wear at all times while in the lab.
- These compounds must be disposed of in an environmentally friendly fashion. After the lab is completed, dispose of them in a departmentally approved waste container. Do not pour them down the sink.

Introduction

Let's introduce a concept, molecular polarity, which chemists frequently use. Polar molecules are attracted to an electric field. On the other hand, nonpolar molecules are unaffected by an electric field. This experiment is based upon the actions of polar and nonpolar molecules in electric fields. Interestingly, molecular polarity also has another easily measurable effect. To a large degree, molecular polarity determines whether or not chemical compounds are soluble in one another as we shall examine in the second half of the experiment.

Being good scientists, chemists want to understand what makes a molecule be polar or nonpolar at the molecular and electronic level. Molecular polarity is a function of two components of molecules, electron distribution and molecular symmetry. Let's look at these two components.

Electrons on molecules are not always evenly distributed across the molecule. If the molecule contains highly electronegative elements (halogen, nitrogen, or oxygen atoms, for example), electrons can be concentrated in that area of the molecule. Concentration of electrons in one area of a molecule increases the electron density (number of electrons per volume) in that area. The overall effect of high electron density is to generate a region of localized negative charge on the molecule near the electronegative atoms. Electrons that are concentrated in one portion of the molecule have been removed from other regions, which decreases the electron density in that part of the molecule. This generates a region of localized positive charge. Chemists designate the localized positive and negative areas on molecules using the symbols δ^+ and δ^-, indicating the existence of partially positive and partially negative charges, respectively, on those parts of the molecule. As shown below, carbon monoxide and carbon dioxide are two simple molecules having electropositive and electronegative regions indicated with the δ^+ and δ^- symbols. In both molecules the highly electronegative O atoms are draining electron density away from the C atom.

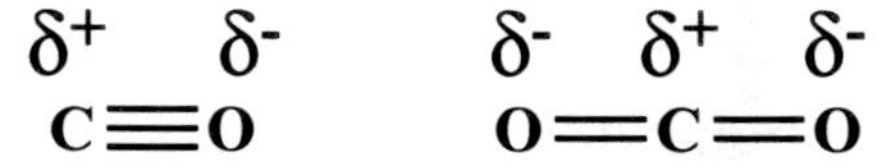

Molecules containing atoms that have small or zero differences in electronegativity, like H_2 or O_2, have equivalent electron distributions across the molecule, making them nonpolar.

Returning to carbon monoxide and carbon dioxide, we can examine the other necessary component of molecular polarity: symmetry. Symmetry refers to whether or not one side of an object corresponds in shape to the other side of the object. For example, humans are bilaterally symmetric. If we pass a plane from head to foot through a human, as indicated below, the left side of the body has the same shape and placement of body parts as the right side.

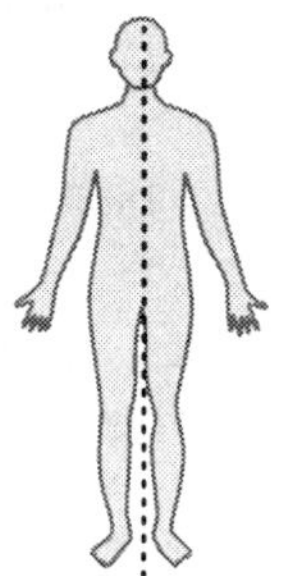

However, if we pass a plane across the center of our bodies, like below, we see that the top of our bodies is not symmetric with the bottom. Humans are symmetric with respect to the first plane and asymmetric with respect to the other.

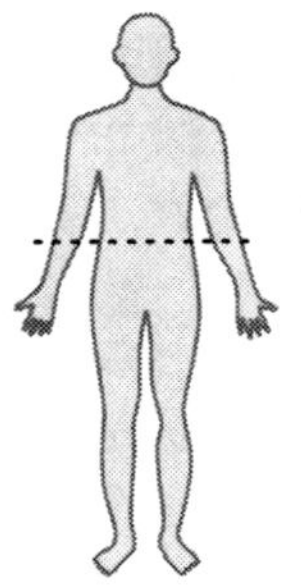

Likewise, molecules can be symmetric with respect to one plane and asymmetric to other planes. Carbon dioxide is symmetric with respect to two planes as shown below.

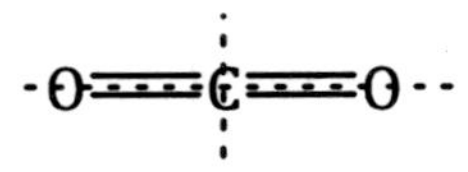

symmetric with respect
to both planes

Carbon monoxide, however, is symmetric with respect to one plane and asymmetric with respect to a second plane (see below).

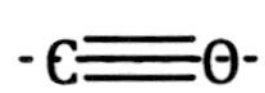

symmetric

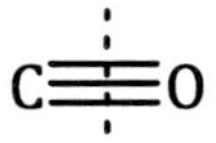

asymmetric

If we then recognize that the asymmetric plane also divides the molecule with respect to its partially positive and negative ends, it is obvious that carbon monoxide has a property that carbon dioxide does not. Carbon monoxide is asymmetric with respect to both atom position and charge distribution. Molecules that have both these properties are polar. Carbon dioxide is nonpolar. Carbon monoxide is polar. Electrons in carbon monoxide reside more frequently near the O atom than near the C atom, generating a measurable electric and magnetic field on the molecule.

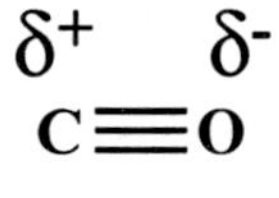

Let's look at two other simple possibilities, BF_3 and NH_3. BF_3 is a planar molecule while NH_3 is trigonal pyramidal. This makes a significant difference in their symmetry. Because it is planar, BF_3 is symmetric. However, because NH_3 is trigonal pyramidal, it is asymmetric to both atom position and charge distribution. BF_3 is nonpolar and NH_3 is polar.

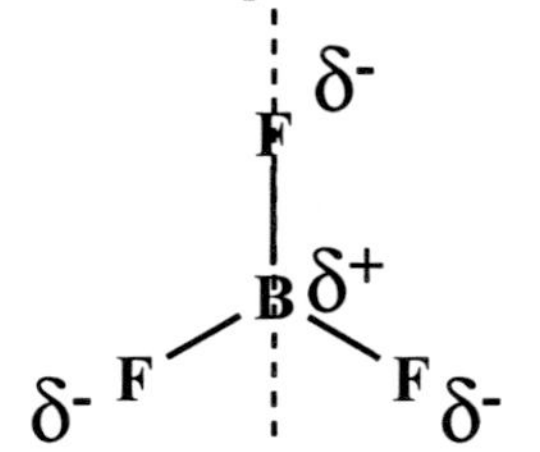

symmetric

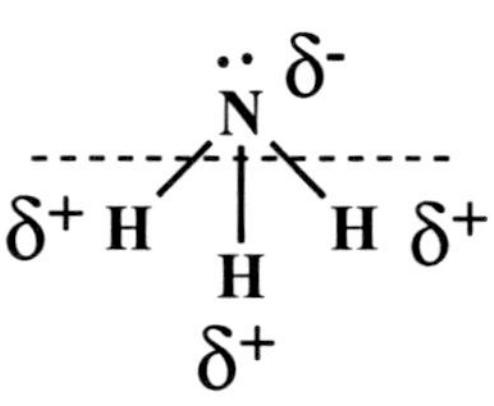

asymmetric

Because they are polar, in the presence of an electric field ammonia (NH_3) molecules align themselves as shown below. In this experiment you will see the effect of this alignment in the presence of the static electric field formed on the surface of a balloon.

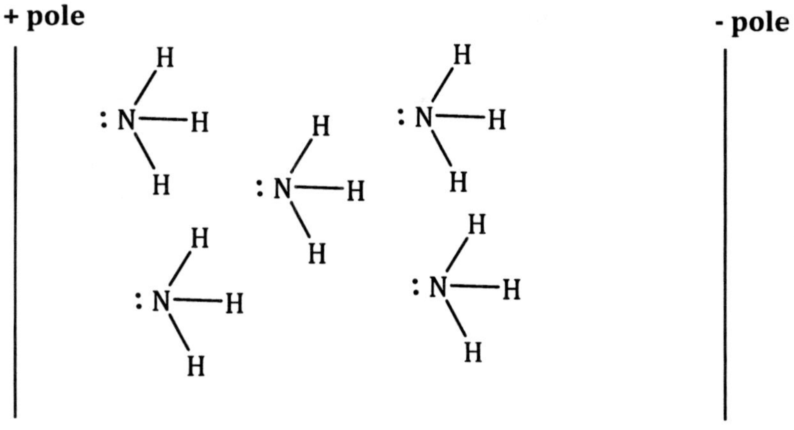

Figure 1. Representation of the Interaction of Polar Ammonia Molecules with an Electric Field

Liquid solubility is also affected by molecular polarity. Liquids composed of polar molecules are soluble in other liquids made of polar molecules. Nonpolar molecular liquids are soluble in other nonpolar molecular liquids. However, polar liquids and nonpolar liquids are insoluble in one another. The basic cause of these solubility differences is that the forces between molecules, intermolecular forces, are similar when polar molecules are mixed with other polar molecules or when nonpolar molecules are mixed with nonpolar molecules. But when nonpolar molecules are mixed with polar molecules the intermolecular forces are so dissimilar that the molecules do not attract one another and do not dissolve. Water is a polar molecule. Cooking oil is a mixture of nonpolar molecules. Cooking oil does not dissolve in water because of the two liquids' intermolecular force differences.

Procedure

Part A – Examining Polarity with an Electric Field

1. Several of the compounds used in this experiment are volatile and flammable. For the sake of safety and efficiency either the lab assistants or you should place seven ring stands with attached buret clamps and burets inside a fume hood or well ventilated work space.
2. You must label each buret with the name of one of the seven compounds used in the experiment: water (H_2O), carbon disulfide (CS_2), dichloromethane (CH_2Cl_2), ethanol (C_2H_5OH), 2-propanone or acetone (CH_3COCH_3), hexane (C_6H_{14}), and ortho-dichlorobenzene ($C_6H_4Cl_2$).
3. Fill each buret with the liquid indicated by the label.
4. Place one of your beakers labeled as water (H_2O) underneath the buret containing water.

5. Blow up a standard balloon and tie it closed. Rub the balloon several times with a piece of wool or fur or across your hair until it has a significant static electric charge.
6. Open the buret stopcock until a steady stream of water is flowing out of the buret into the labeled beaker.
7. Hold the charged balloon near but not in the stream of water, as shown in **Figure 2**. Observe what happens to the water stream as the balloon approaches. **Water is a polar molecule**.

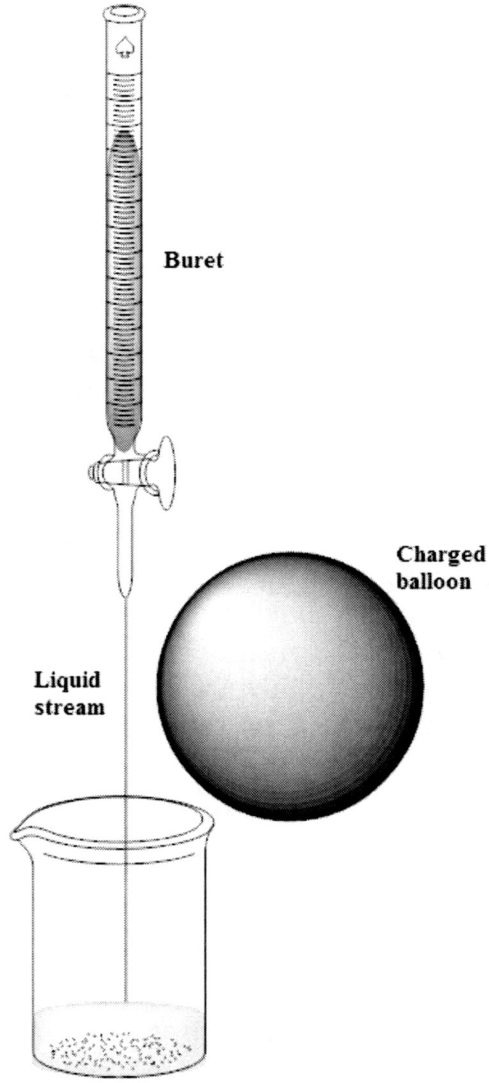

Figure 2. Experimental Setup

8. Repeat the procedure, holding the charged balloon near the carbon disulfide (CS_2) buret stream. Catch the carbon disulfide in a different labeled beaker. You may need to recharge the balloon by rubbing it again with wool, fur, or on your hair. If the balloon touches the carbon disulfide stream, you must dispose of the

balloon, inflate another one, and repeat the charging procedure before holding the balloon near the stream. **Do not rub a balloon that has touched carbon disulfide (CS_2) on your hair**.

9. Observe what happens as the balloon approaches the CS_2 stream. Based upon your observation of how the polar water stream was affected by the balloon and how carbon disulfide is affected, can you conclude that carbon disulfide is polar or nonpolar?
10. Repeat the procedure, holding the charged balloon near the dichloromethane (CH_2Cl_2) buret stream. Catch the dichloromethane in a different labeled beaker. Recharge the balloon if needed. As before, if the balloon contacts the dichloromethane you must discard it and inflate another balloon. **Do not rub a balloon that has touched dichloromethane (CH_2Cl_2) on your hair**. Based upon your observation of how the polar water stream was affected by the balloon and how dichloromethane is affected, can you conclude that dichloromethane is polar or nonpolar?
11. Repeat this procedure with the burets containing ethanol (C_2H_5OH), 2-propanone or acetone (CH_3COCH_3), hexane (C_6H_{14}), and ortho-dichlorobenzene ($C_6H_4Cl_2$). Catch each liquid in a different labeled beaker. Based upon their behaviors determine if each compound is polar or nonpolar.

Part B – Effect of Polarity on Solubility

12. Pour from the beaker labeled water 5 to 10 mL of water into a disposable test tube. Then pour from the beaker labeled carbon disulfide 3 to 5 mL of carbon disulfide (CS_2) into the test tube containing the water. Are the two liquids soluble or insoluble in each other? Do the liquids separate into two layers or is there just one layer? Does this observation confirm what you learned from the balloon experiment?
13. You must safely dispose of the carbon disulfide water mixture as directed by your instructor.
14. Using another disposable test tube, repeat the above procedure using water and dichloromethane, CH_2Cl_2. Again note if the two liquids are soluble or insoluble in each other.
15. Repeat this procedure, in turn, to determine the solubilities of ethanol (C_2H_5OH), 2-propanone or acetone (CH_3COCH_3), hexane (C_6H_{14}), and ortho-dichlorobenzene ($C_6H_4Cl_2$) in water.
16. Repeat the solubility tests by pouring 5 to 10 mL of hexane into a clean disposable test tube then adding 3 to 5 mL of CS_2. Note if the two liquids are soluble or insoluble in each other.
17. After safely disposing of the mixture, repeat the solubility tests in hexane using, in turn, CH_2Cl_2, C_2H_5OH, CH_3COCH_3, and $C_6H_4Cl_2$.

Experiment 9 Pre-Lab (Show ALL Work and ALWAYS include units)

1. Water is a polar molecule. Shown below is the Lewis structure of water. Draw on this structure where the partially positive and negative parts of the molecule are. Draw planes through the molecule that are symmetric and asymmetric with respect to both atom position and charge distribution.

2. Shown below are the Lewis structures of three different dichlorobenzenes. Draw on these structures where the partially positive and negative parts of the molecule are. Draw planes through each molecule that are either symmetric or asymmetric with respect to both atom position and charge distribution. Based upon your drawings, decide if the molecules are polar or nonpolar.

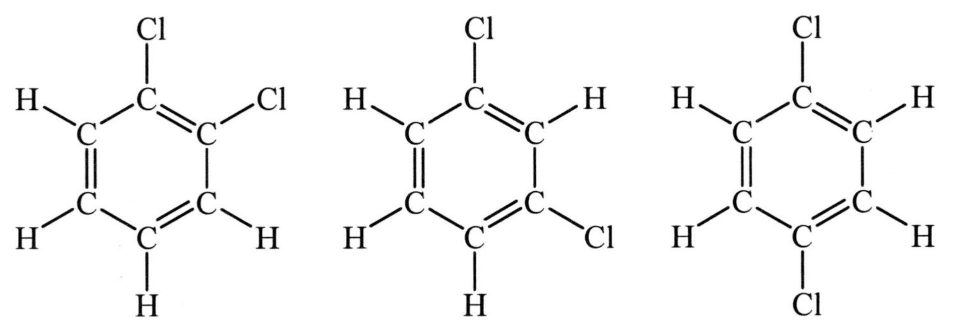

3. Isomers are chemical compounds that have the same molecular formula but differ in atom placement. Suppose that a square planar molecule exists having the formula XH_2Cl_2, where X is an undesignated element that is the central atom in the molecule. Two isomers of this compound would exist as indicated in the drawing below. Are the isomers polar or nonpolar? Designate each as either polar or nonpolar.

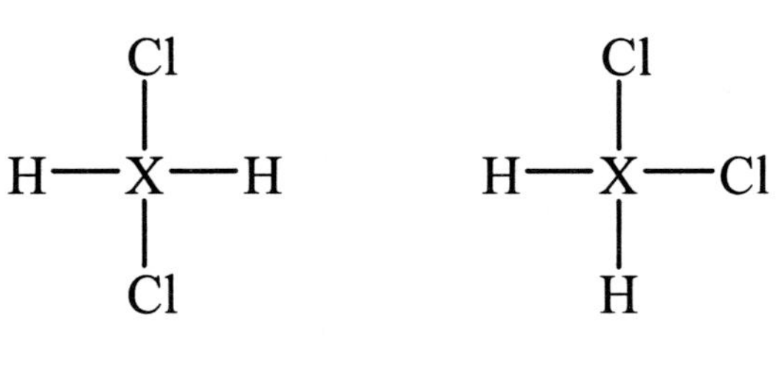

4. Suppose that a tetrahedral molecule exists having the formula XH_2Cl_2, where X is an undesignated element that is the central atom in the molecule. Two different drawings of this compound are shown below. Are these isomers of one another? Explain your logic in answering this question. Designate each as either polar or nonpolar. (You may want to borrow a molecular model kit from your instructor to answer this question correctly.)

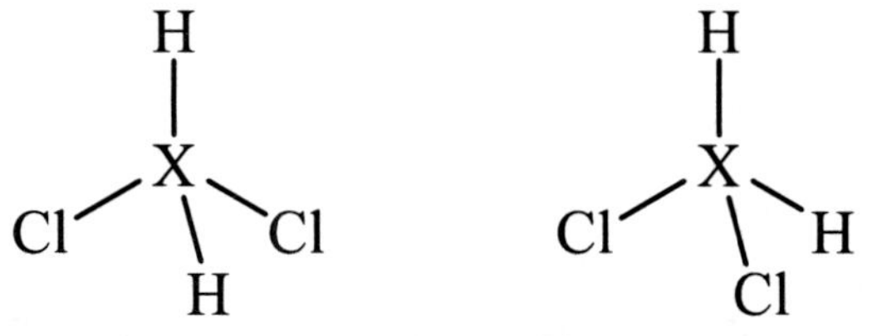

Experiment 9 Lab Report (Show ALL Work and ALWAYS include units)

Part A – Examining Polarity with an Electric Field

1. Describe in your own words what happened when the charged balloon came close to the stream of water from the buret tip.

2. Describe in your own words what happened when the charged balloon came close to the stream of carbon disulfide, CS_2, from the buret tip. Knowing that water is polar, does this behavior indicate that CS_2 is polar or nonpolar? Draw a Lewis structure of CS_2 that conforms to your experimental observation.

3. Describe in your own words what happened when the charged balloon came close to the stream of dichloromethane, CH_2Cl_2, from the buret tip. Knowing that water is polar, does this behavior indicate that CH_2Cl_2 is polar or nonpolar? Other experiments have shown that CH_2Cl_2 does not have any isomers. Based upon the prelab exercise and your observation from this experiment draw a Lewis structure of CH_2Cl_2.

4. Describe in your own words what happened when the charged balloon came close to the stream of ethanol, C_2H_5OH, from the buret tip. Knowing that water is polar, does this behavior indicate that C_2H_5OH is polar or nonpolar? Draw a Lewis structure of C_2H_5OH that conforms to your experimental observation.

5. Describe in your own words what happened when the charged balloon came close to the stream of 2-propanone or acetone, CH_3COCH_3, from the buret tip. Knowing that water is polar, does this behavior indicate that CH_3COCH_3 is polar or nonpolar? Shown below is a skeletal structure of CH_3COCH_3. Complete a Lewis structure of CH_3COCH_3 on the skeletal structure that conforms to your experimental observation.

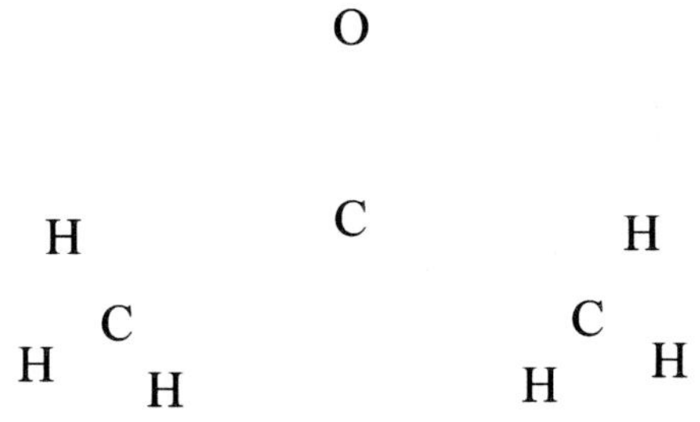

6. Describe in your own words what happened when the charged balloon came close to the stream of hexane, C_6H_{14}, from the buret tip. Knowing that water is polar, does this behavior indicate that C_6H_{14} is polar or nonpolar? Shown below is a skeletal structure of C_6H_{14}. Complete a Lewis structure of C_6H_{14} on the skeletal structure that conforms to your experimental observation.

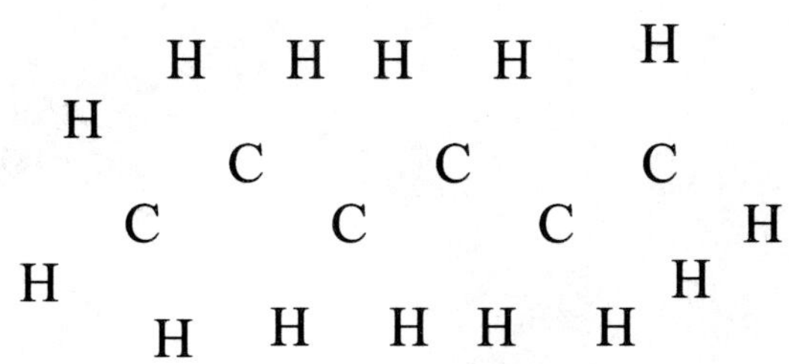

7. Describe in your own words what happened when the charged balloon came close to the stream of ortho-dichlorobenzene, $C_6H_4Cl_2$, from the buret tip. Knowing that water is polar, does this behavior indicate that $C_6H_4Cl_2$ is polar or nonpolar? One of the structures shown below is ortho-dichlorobenzene. Based upon your observation, which one of the two structures is ortho-dichlorobenzene?

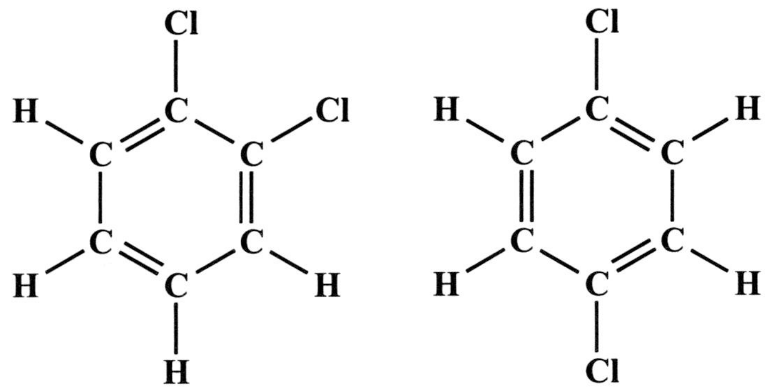

Part B – Effect of Polarity on Solubility

8. Complete the following table by writing soluble or insoluble in the provided spaces based upon your balloon and solubility observations.

Solute Compounds	Polar or Nonpolar Compound	Solvent Water (H_2O)	Solvent Hexane (C_6H_{14})
Carbon disulfide CS_2			
Dichloromethane, CH_2Cl_2			
Ethanol, C_2H_5OH			
2-propanone, CH_3COCH_3			
Hexane, C_6H_{14}			
Ortho-dichlorobenzene, $C_6H_4Cl_2$			

9. Is there a solubility pattern that you can discern from the data presented above? If so, describe that pattern. If not, state that there is no pattern.

Experiment 10
Thermal Gravimetric Analysis –
Analyzing the Composition of a Mixture by Heating It

Objectives

- Use stoichiometry as a tool.
- Understand the basis of thermal gravimetric analysis.

Topics Introduced in this Lab

- Reaction stoichiometry
- Mixture separation
- Thermal gravimetric analysis

Related Zumdahl sections

- Chapter 5 Sections 1, 4, and 5

Safety Considerations

- We will heat the mixture to temperatures greater than 200 °C. The crucible and hot plate will be hot to the touch. If your skin contacts either one you will experience a very serious burn. Handle with care.
- 6 *M* HCl(*aq*) is used to clean the crucible. This is both a strong and concentrated acid that can cause severe skin burns. HCl in eyes results in serious eye damage. Be very careful with this substance. If any gets on your skin or eyes wash with copious amounts of water.
- Do not ingest any of these compounds.
- Wear safety glasses and appropriate lab wear at all times while in the lab.

Introduction

Pure substances are seldom found in nature. Instead, mixtures of various substances are frequently encountered. Many techniques are known that can be used to separate substances in a mixture. In some cases simple filtration is all that is required. If the mixture contains substances that are soluble in different solvents, a method called extraction is commonly used. However, in some cases it is difficult to separate a mixture. In these instances, we may still determine the mixture composition if the components respond to heat differently. For a mixture of solids, if one component decomposes in a known fashion upon heating while the others do not it is easy to determine the component amounts using thermal gravimetric analysis.

Potassium chloride, KCl, and sodium hydrogen carbonate, $NaHCO_3$, have very different thermal stabilities, and therefore respond differently to being heated to a temperature of 200 to 250 °C. KCl does not decompose or react at these temperatures. However, $NaHCO_3$ readily decomposes into sodium carbonate, gaseous water, and carbon dioxide at these temperatures (see **Equation1**).

$$2NaHCO_3(s) \rightarrow Na_2CO_3(s) + H_2O(g) + CO_2(g)$$

Equation 1. $NaHCO_3$ Decomposition Reaction

A solid mixture composed of KCl and $NaHCO_3$ will lose mass when heated sufficiently. All of the mass loss is due to the decomposition of $NaHCO_3$ and none from the KCl. From the decomposition stoichiometry, we can determine the mass loss of water and carbon dioxide for every gram of sodium hydrogen carbonate that decomposes in the mixture. If we determine the mass of the mixture before and after heating, the mass difference allows us to calculate the mass of sodium hydrogen carbonate in the original mixture and therefore the percentage of each component ($NaHCO_3$ and KCl).

To ensure that all of the water and carbon dioxide have been removed it is necessary to heat the mixture more than once. After each heating, the mixture and its container (called a crucible) are weighed until the mass differs by less than 0.001 g. It is important that the crucible containing the mixture be at room temperature before weighing it. Hot crucibles in the weighing area of an analytical balance establish convection currents in air that affect the measured crucible mass. Therefore, you MUST allow the crucible and its contents to cool to room temperature after each heating cycle and before weighing.

Procedure

Part A – Clean and Weigh the Crucible

1. Obtain a crucible and lid. To ensure that the crucible is clean place 2 to 3 mL of 6 *M* HCl(*aq*) into the crucible for 2 to 3 minutes. Be careful with HCl(*aq*) as a few drops of it on your skin can result in severe burns.
2. Pour the HCl(*aq*) into a waste jar and then wash the crucible with lots of water to remove any HCl(*aq*) residue. Be careful while washing the crucible as the HCl(*aq*) residue could still react with your skin or eyes.
3. Completely dry the crucible and top.
4. Place the crucible onto the hot plate. Place the top onto the crucible slightly ajar so that moisture can escape, **Figure 1**.
5. Turn the heating plate to 150 °C and allow the crucible to reach that temperature for at least 5 minutes. This will ensure that the crucible is completely dry.
6. Using crucible tongs, place the top securely on the crucible and then remove the crucible and top from the heating plate*. Place both onto a wire gauze to allow them to cool to room temperature. This will take 10 to 15 minutes. Holding your hand over but not in contact with the crucible is a good way to determine if the crucible has cooled.
   *If you are using a combination hot/stir plate, the magnetic stirrer can exert a slight pull on the metal crucible tongs.

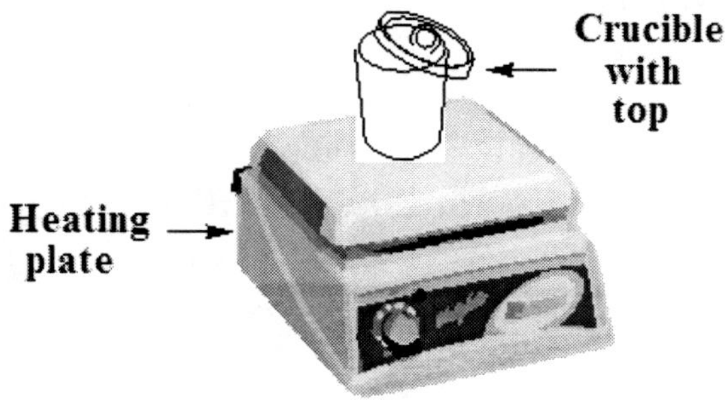

Figure 1. Experimental Setup

7. Once the crucible and top have cooled to room temperature use crucible tongs to take them to the balance area. Do not touch the crucible as fingerprints have enough mass to be detected by an analytical balance. Weigh and record the mass of the crucible and top.

Part B – Determine Mass Loss from 1st Heating

8. Place approximately 1 g of the $KCl/NaHCO_3$ mixture into the crucible. Weigh and record the mass of the crucible, top and mixture. Determine the mixture mass by difference.
9. Place the crucible containing the mixture back onto the heating plate. Set the crucible top on the crucible slightly ajar again. Heat the crucible to 150^{0}C for five minutes.
10. Turn up the heating plate from 150 °C to 200 °C. Heat the crucible at 200 °C for another 5 minutes.
11. Place the top securely on the crucible. Remove the crucible and top from the heating plate and place them on the wire gauze to cool.
12. Once the crucible and top have cooled to room temperature, take them to the weighing area using crucible tongs. Weigh and record the mass of the crucible, top, and remaining mixture.
13. Repeat steps 9 through 12.
14. If the crucible mass is within ±0.001 g of the mass determined in step 12, the experiment is completed. If the mass is more than ±0.001 g different than that of step 12, repeat steps 10 through 12.
15. Repeat step 14 until the crucible and mixture combination achieves a constant mass.

Part C – Determine Mass Loss from 2nd Heating

16. Repeat steps 1 to 15 with a second mixture sample.

Experiment 10 Pre-Lab (Show ALL Work and ALWAYS include units)

1. From the decomposition reaction of sodium hydrogen carbonate determine the mass of $H_2O(g)$ lost per gram of $NaHCO_3(s)$ heated.

$$2NaHCO_3(s) \rightarrow Na_2CO_3(s) + H_2O(g) + CO_2(g)$$

2. From the decomposition reaction of sodium hydrogen carbonate determine the mass of $CO_2(g)$ lost per gram of $NaHCO_3(s)$ heated.

$$2NaHCO_3(s) \rightarrow Na_2CO_3(s) + H_2O(g) + CO_2(g)$$

3. What is the total mass loss per gram of $NaHCO_3(s)$ heated?

4. 2.350 g of a KCl and $NaHCO_3$ mixture are heated to constant mass. When cooled and weighed the remaining mixture had a mass of 1.950 g. Using this data determine the percentages of KCl and $NaHCO_3$ in the original mixture.

Experiment 10 Lab Report (Show ALL Work and ALWAYS include units)

Part A – Clean and Weigh Crucible

1. Mass of Clean Crucible and Top ________

Part B – Determine Mass Loss from 1st Heating

2. Mass of Crucible, Top, and Mixture ________
3. Mass of Crucible, Top, and Mixture After 1st Heating ________
4. Mass of Crucible, Top, and Mixture After 2nd Heating ________
5. Mass of Crucible, Top, and Mixture After 3rd Heating ________
6. Mass of Crucible, Top, and Mixture After 4th Heating ________
7. If more than 4 heatings are required record those values below.

Part C – Determine Mass Loss from 2nd Heating

8. Mass of Crucible, Top, and Mixture ________
9. Mass of Crucible, Top, and Mixture After 1st Heating ________
10. Mass of Crucible, Top, and Mixture After 2nd Heating ________
11. Mass of Crucible, Top, and Mixture After 3rd Heating ________
12. Mass of Crucible, Top, and Mixture After 4th Heating ________
13. If more than 4 heatings are required record those values below.

14. Determine the mass of the initial mixture placed into the crucible.

15. Determine the mixture mass loss after the first heating.

16. Determine the total mixture mass loss after the second heating.

17. Determine the total mixture mass loss after the third heating, if necessary.

18. Determine the total mixture mass loss after the fourth and subsequent heatings, if necessary.

19. Using the total mixture mass loss (after heating to constant mass) calculate the original amount of $NaHCO_3$ present in the mixture.

20. Calculate the percentage of $NaHCO_3$ and KCl in the original mixture.

Experiment 11
Understanding Limiting Reagents

Objectives

- Learn about the concept of limiting reagents by performing an experiment with visual evidence of a limiting reagent.
- Practice with molarity and stoichiometry calculations.

Topics Introduced in this Lab

- Limiting reagent
- Molarity
- Sneak peek at the ideal gas law

Related Zumdahl Sections

- Chapter 5 Section 11
- Chapter 6 Section 3
- Chapter 8 Section 3

Safety Considerations

- HCl is corrosive. If you get any on you, wash immediately with copious amounts of water followed by washing with soap and water.
- The reactions in this experiment produce H_2 gas. H_2 gas is highly flammable. No open flames should be present in the lab during this experiment.
- Magnesium powder is also highly flammable.
- Wear safety glasses and appropriate lab wear at all times while in the lab.

Introduction

As you've already seen, chemical reaction stoichiometry is related to the stoichiometric coefficients in the balanced chemical equation. Reaction stoichiometry tells us the relative amounts of reactants that are consumed in a chemical reaction and the relative amounts of products formed. Reactants *must* react in the ratio dictated by the reaction stoichiometry. If one of the reactants is in short supply relative to the others according to the reaction stoichiometry, then that reactant dictates the amount of product that can be made and is the limiting reagent.

In this experiment, you will examine the reaction of hydrochloric acid with magnesium metal as illustrated by the following balanced chemical equation, **Equation 1.**

$$\mathbf{2HCl}(aq) + \mathbf{Mg}(s) \rightarrow \mathbf{MgCl_2}(s) + \mathbf{H_2}(g)$$

Equation 1. Equation for Reaction of HCl(*aq*) with Mg(*s*)

From the balanced equation, we know that 2 moles of HCl react with 1 mole of Mg to produce 1 mole of $MgCl_2$ and 1 mole of H_2. This means that every time the reaction occurs there must be *exactly* twice as many moles of HCl as Mg. If there are more than twice as many moles of HCl as Mg, then all the Mg is consumed first and is the limiting reagent. If there are less than 2 times as many moles of HCl as Mg, then HCl is consumed first and is the limiting reagent.

For this experiment, you will perform a series of tests in which the amount of HCl is kept constant while the amount of Mg metal is varied. For each test, you will collect the H_2 gas produced by the reaction and use the gas volume to determine the moles of gas produced. Based upon the moles of gas produced in each test, you will determine the limiting reagent.

The HCl used in this experiment is dissolved in water, an aqueous solution. To determine the moles of HCl used in the reaction, you must know the HCl solution concentration and use that concentration to determine the moles contained in a specified volume. A common solution concentration unit is molarity (*M*). Molarity is defined as shown in **Equation 2.**

$$\textbf{Molarity} = \frac{\textbf{moles of solute}}{\textbf{liters of solution}}$$

Equation 2. Molarity

It is important to note that liters of solution includes the total volume of the solution, both the volume of the solute AND the solvent. If the solution molarity is known, the moles of solute in a given volume can be calculated. For instance, in 20.0 mL of a 1.0 *M* HCl solution, there are 0.020 moles of HCl, **Equation 3.**

Keeping in mind that 20.0 mL = 0.0200 L,

$$\mathbf{(1.0\ \mathit{M}) \times (0.0200\ L) = 0.020\ moles\ of\ HCl}$$

Equation 3. Determination of Moles from Molarity and Volume

Since you are collecting H_2 gas in this experiment to determine the moles of product, you must convert from gas volume to moles of gas. To do this, you will use the ideal gas law (shown in **Equation 4**).

$$PV = nRT$$

where P **= gas pressure in atmospheres (atm)**

V **= gas volume in liters (L)**

n **= moles of gas**

R **= universal gas constant =** $0.08206 \dfrac{\text{L}\cdot\text{atm}}{\text{mol}\cdot\text{K}}$

T **= temperature in kelvin (K)**

Equation 4. Ideal Gas Law with Variables Defined

From the ideal gas law, you can see that from the pressure, volume, and temperature of a gas, you can calculate the number of moles of gas as shown in **Equation 5.**

$$n = \frac{PV}{RT}$$

Equation 5. Calculation of Moles Using the Ideal Gas Law

Using the ideal gas law allows us to calculate the moles of gas produced for each test.

Procedure

Part A – Collecting Hydrogen Gas

1. Gather the following equipment:
   - 125-mL filter flask and rubber stopper
   - Piece of tubing (18-24 inches long)
   - 100-mL graduated cylinder
   - 50-mL graduated cylinder
   - 600-mL beaker
   - 2 ringstands and 2 utility clamps
2. Using a graduated cylinder, add exactly 40.0 mL of 0.100 *M* HCl solution to your flask and place a rubber stopper in the top of the flask.
3. Using weigh paper, weigh out exactly 0.0243 grams of magnesium powder. Place this to the side for the moment.
4. Fill a 100-mL graduated cylinder to the top with water. Place a 600-mL beaker upside down on the top of the graduated cylinder (see Figure 1A). Holding the beaker and graduated cylinder together, flip the assembly so that the beaker is now upright and the graduated cylinder is upside down. Immediately pour approximately 300 mL of water into the beaker. You should be able to lift the graduated cylinder slightly off of the bottom of the beaker without the water level in the cylinder decreasing significantly. Place the beaker/graduated cylinder assembly on the bench top and secure the graduated cylinder approximately 1 inch from the bottom of the beaker using a ring stand and a utility clamp (see Figure 1B).

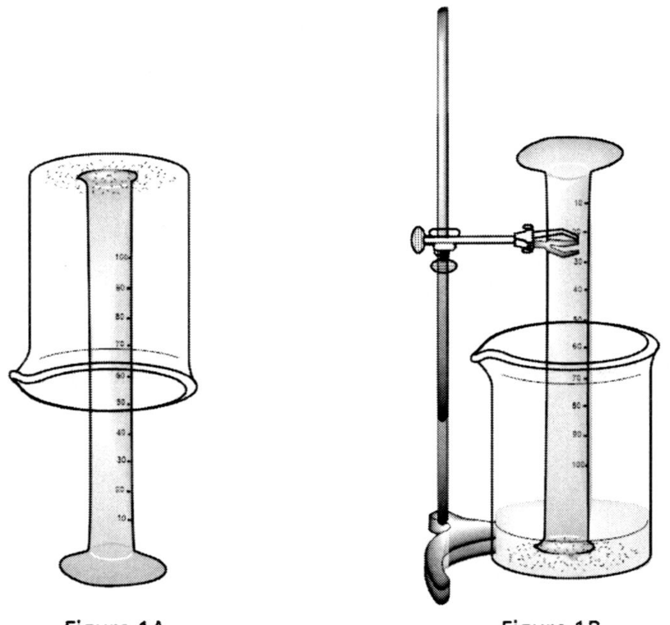

Figure 1. Setting up the Gas Collection Apparatus

5. Attach one end of the tubing to the filter flask. At this time, you should use an additional ring stand and utility clamp to secure the flask and ensure that it does not tip over.
6. Fix the other end of the tubing so that it is in the beaker of water and inside the graduated cylinder. It is okay if some water gets into the tubing. The final experimental setup should resemble that shown in Figure 2.

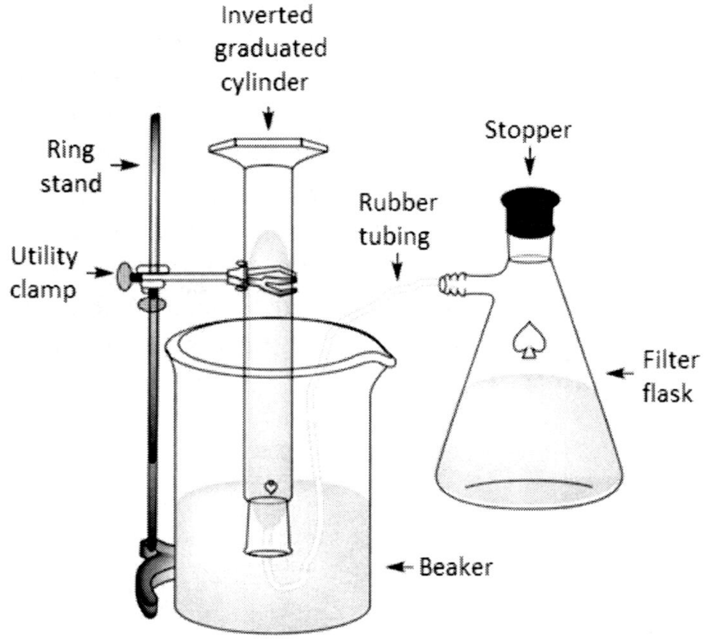

Figure 2. Final Experimental Setup

7. Record the initial volume of water in the graduated cylinder, noting that the volume markings are inverted.
8. Unclamp your filter flask. With the rubber stopper in hand, add the magnesium powder to your flask and immediately re-cap the flask with the stopper. Swirl the flask while the reaction occurs.
9. Notice that as gas is produced, it flows into the graduated cylinder, displacing the water and causing the water level to decrease.
10. Wait until the water level no longer changes. At that time, stop swirling the flask, re-secure it so that it does not tip over, and record the final water volume in the graduated cylinder. Determine the volume of gas produced by the reaction of magnesium with HCl. Record this volume in your lab report.
11. Measure the temperature of the water in the beaker and record it in your lab manual. We can assume that the temperature of the water equals the collected gas temperature.
12. Ask your instructor for today's measured atmospheric pressure. We will assume that the pressure of the collected gas equals atmospheric pressure. Record this value in your lab manual.
13. Empty the contents of the flask into a waste container and rinse the flask thoroughly with water before using it again.
14. Repeat steps 2-12 to perform two more trials, using 0.0486 g of magnesium for the second trial, and 0.0729 grams of magnesium for the third trial. Use 40.0 mL of 0.100 *M* HCl solution for each trial.

Experiment 11 Pre-Lab (Show ALL Work and ALWAYS include units)

1. How many moles of HCl are present in 3.0 mL of aqueous HCl solution that is 0.050 M?

2. How many moles of magnesium are present in 0.153 grams of magnesium metal?

3. Suppose you collect some H_2 gas from a reaction. The gas occupies a volume of 20.0 mL. You determine that the temperature of the gas is 25.3 °C and that the gas has a pressure of 640.0 torr. Use the ideal gas law ($PV = nRT$) to determine the moles of H_2 gas collected. Be careful with your units!

4. If you react 0.24 moles of HCl with 0.18 moles Mg, which reactant is the limiting reactant and why?

Experiment 11 Lab Report (Show ALL Work and ALWAYS include units)

Part A – Collecting Hydrogen Gas

1. Fill out the following table.

Trial #	Volume of 0.100 *M* HCl used	Mass of Mg used	Initial water level in grad. cylinder	Water level in grad. cylinder after collecting gas	Volume of H_2 gas collected	Temp (°C) of water/gas
1						
2						
3						

2. Pressure of collected gas (atmospheric pressure): ______________

3. Fill out the following table. Show the appropriate calculations below the table.

Trial #	Moles of HCl used	Moles of Mg used	Moles of H_2 collected	Expected moles of H_2 based on moles of HCl used	Expected moles of H_2 based on moles of Mg used	Identity of limiting reagent (HCl or Mg or Neither)
1						
2						
3						

Moles of HCl calculations:

Trial 1:

Trial 2:

Trial 3:

Moles of Mg calculations:

Trial 1:

Trial 2:

Trial 3:

Moles of $H_2(g)$ collected calculations (use $PV = nRT$):

Trial 1:

Trial 2:

Trial 3:

4. Explain how you determined which reactant was limiting in question #3 above for each trial. Be detailed in your explanations.

   Trial 1:

   Trial 2:

   Trial 3:

Experiment 12
Solution Concentration and Titrations

Objectives
- Learn how to make solutions of specific concentrations.
- Perform an acid-base reaction.
- Learn how to perform a titration and use the data to calculate exact concentration (standardization).

Topics Introduced in this Lab
- Molarity
- Acid-base reactions
- Titrations

Related Zumdahl Sections
- Chapter 6, Sections 3 and 8

Safety Considerations
- Safety goggles and appropriate lab attire must be worn at all times when working in the lab.
- Do not ingest any of these chemicals.
- NaOH is extremely corrosive. Avoid handling solid NaOH pellets. If you touch the solid pellets or you spill NaOH solution on yourself, rinse immediately with copious amounts of water followed by washing with soap and water.

Introduction
Solutions of chemical compounds are made by dissolving the species of interest (the solute) in another chemical compound. This second compound, frequently a liquid, is called the solvent. Solutions are used widely in chemistry because they allow chemical species to easily approach one another so that they can undergo a chemical reaction. To successfully use solutions to run chemical reactions, it is essential to quantify the amount of a chemical species in a given volume of solution. Therefore, a unit of concentration is needed. The most common unit of concentration used in chemistry is molarity (*M*), which is defined as the moles of solute per liter of solution, **Equation 1**. It is important to recognize that this is liters of solution and NOT liters of solvent.

$$M = \frac{\textbf{moles of solute}}{\textbf{L of solution}}$$

Equation 1. Molarity

Simply understanding what molarity means is not enough. As chemists, we want to make solutions of specific concentration to then use in chemical reactions. So how do we make solutions of a specific concentration? From the equation for molarity we can see that there are two quantities that need to be controlled to create a solution of a given concentration (*M*): the moles of solute and the total volume of the solution. Generally, chemists will first determine what volume of a solution of a certain concentration they want to make depending on what they need it for. Let's say that you wanted to make 100. mL of a 0.25 *M* NaCl solution. From this information, we already know two of the three variables in our molarity equation, which now becomes **Equation 2.**

$$\mathbf{0.25\ \mathit{M} = \frac{moles\ of\ solute}{0.100\ L\ of\ solution}}$$

Equation 2. Molarity for NaCl Solution

We can now easily solve for the moles of solute (**Equation 3**).

$$\mathbf{moles\ of\ solute = \left(0.25\ \frac{mol}{L}\right)(0.100\ L) = 0.025\ mol}$$

Equation 3. Determining Moles of NaCl Needed

From the number of moles, we can calculate how many grams of NaCl to use. The molar mass is used to calculate the grams of NaCl needed in **Equation 4.**

$$\mathbf{0.025\ mol\ NaCl\left(\frac{58.44\ g}{mol\ NaCl}\right) = 1.5\ g\ NaCl\ needed}$$

Equation 4. Determining Mass of NaCl Needed

Great! We now know that 1.5 grams of NaCl dissolved in enough water to give 100. mL of solution will result in a solution that is 0.25 *M*. But how do we make exactly 100. mL of solution? If you dissolve 1.5 grams of NaCl in 100. mL of water, the total volume will be more than 100. mL. The NaCl takes up space after all! To make solutions accurately, chemists use a device called a volumetric flask (**Figure 1**).

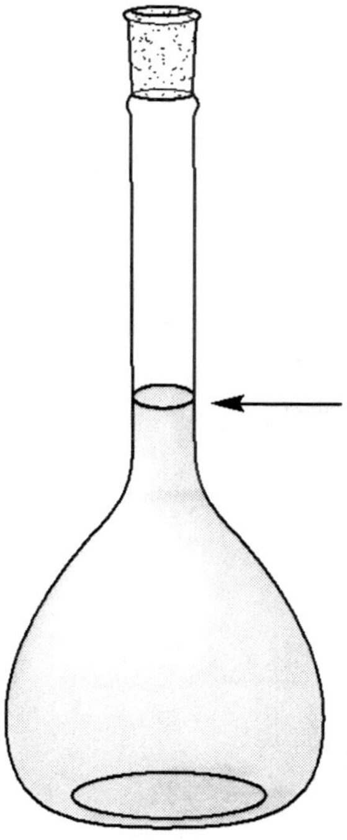

Figure 1. Volumetric Flask

Volumetric flasks have a single mark on the neck of the flask that corresponds to a specific volume. For instance, if a 100-mL volumetric flask is used, when the bottom of the meniscus of the solution in the flask is just at the line on the neck, the flask contains exactly 100.0 mL of solution. The typical procedure for using a clean volumetric flask is as follows:

1. Add some solvent to the flask.
2. Add the necessary amount of solute to the flask.
3. Add a little more solvent and swirl to dissolve all solute.
4. Add solvent until the bottom of the meniscus touches the line on the neck of the flask.
5. Cover the flask and mix the contents thoroughly by swirling and inverting.

Using this procedure creates a solution of a desired volume and concentration while accounting for the volume of the dissolved solute.

Sometimes, chemists need to make a solution of desired concentration, not from solute and solvent as described above, but from a previously prepared solution of higher concentration. This is called a dilution. This is particularly useful when chemical reagents (such as HCl) are sold as concentrated solutions.

When performing a dilution, a specific volume of concentrated solution is combined with more solvent to create a new solution of lower concentration. While this may seem more complicated than our first technique of making a solution, it is still simply a process of obtaining the correct number of moles of solute in the appropriate volume of solution. An example is given in Example 1 to make this point clear.

You are given a solution of HCl dissolved in water that is 6.00 *M* and must make 250.0 mL of a solution that is 1.00 *M* for use in an experiment.

250.0 mL of a 1.00 *M* solution must have 0.250 moles of HCl dissolved in it by definition.

$$\frac{x \text{ moles}}{0.2500 \text{ L}} = 1.00\ M$$
$$x = 0.250 \text{ moles}$$

Therefore, you need to determine what volume of the concentrated solution (6.00 *M*) will give you 0.250 moles of HCl.

$$\frac{0.250 \text{ moles}}{x \text{ L}} = 6.00\ M$$
$$x = 0.0417 \text{ L}$$

0.0417 L or 41.7 mL of the 6.00 *M* solution has 0.250 moles of HCl in it.

To make 250.0 mL of solution that is 1.00 *M* from the 6.00 *M* solution, first add ~100 mL of water to a 250-mL volumetric flask. Next, add 41.7 mL of the 6.00 *M* HCl solution to the flask and swirl to mix. Then, add water until the total solution level reaches the mark on the neck of the flask, cover, and mix thoroughly. Those 0.250 moles of HCl are now dissolved in 250.0 mL of solution, giving a solution that is 1.00 *M*.

Example 1. Example of a Dilution Procedure and Necessary Calculations

The same calculation can be done in a single step as shown in **Example 2.**

$$(M_i)(V_i) = (M_f)(V_f)$$

where M_i = molarity of initial solution
V_i = volume of initial solution
M_f = molarity of final solution
V_f = volume of final solution

$$(6.00\ M)(x\ L) = (1.00\ M)(0.2500 \text{ L})$$
$$x = 0.0417 \text{ L or } 41.7 \text{ mL}$$

Example 2. Dilution Calculations in Single Step Form

In this experiment, you will be asked to create two solutions of NaOH having specific concentrations. You will first make a solution of NaOH using solid NaOH and water and then you will make a second solution by diluting the first. To verify the concentration of the solutions you make (standardize your NaOH solutions), you will perform titrations.

A titration is a technique that involves the reaction of two chemical species, one for which the amount in moles is known accurately, until they reach their stoichiometric equivalence. Knowing the reaction's stoichiometric ratio from the balanced chemical equation and the moles of one of the reagents that was used, the moles of the other reagent can be determined. If the volume of each reagent's solution used is also known, then molarity can be calculated. To illustrate this, imagine that you are mixing A and B and that they react according to the balanced chemical equation shown in **Equation 5**.

$$\mathrm{A + 2B \rightarrow AB_2}$$

Equation 5. Balanced Equation for the Reaction of A with B

Now imagine that you added a 0.10 *M* solution of A to 10.0 mL of a solution of B in a beaker until the exact point where all of B had reacted. No more A was added beyond this point. If you determined that it took 7.0 mL of the 0.10 *M* solution of A to react with all of the B in that 10.0 mL, then you can determine the moles of B that must have been present in that 10.0 mL, and therefore the concentration of that solution of B. This is illustrated in **Example 3** below. In this example, we would say that solution B was titrated using solution A.

$$\left(\frac{0.10\ \text{mol}}{\text{L}}\right)(0.0070\ \text{L}) = 0.00070\ \text{mol A used}$$

$$(0.00070\ \text{mol A})\left(\frac{2\ \text{mol B}}{1\ \text{mol A}}\right) = 0.0014\ \text{mol B in beaker}$$

$$\left(\frac{0.0014\ \text{mol B}}{0.0100\ \text{L}}\right) = 0.14\ M\ \text{concentration of B}$$

Example 3. Determining Concentration from a Titration

The point at which the stoichiometric equivalence of two reactants occurs in a titration is called the *equivalence point*. So how do we determine when we've reached the equivalence point? If the reactants are an acid (proton donor) and base (proton acceptor), the H^+ concentration changes as the reaction progresses. This is shown in **Equation 6**. pH is a simplified way of expressing H^+ concentration and is defined as the negative log of the H^+ concentration in a solution (**Equation 7**). Because of this definition, pH increases as the H^+ concentration decreases. A change in H^+ concentration necessarily leads to a solution pH change. As the equivalence point is approached in an acid-base titration, the pH undergoes a drastic change. A plot of pH versus volume of added base is shown in **Figure 4**. This plot is the *titration curve* that would result if base was slowly added to an acid solution as the pH is monitored. Notice that the equivalence point occurs at a sharp change in the pH. If a titration curve like this can be produced, the equivalence point can therefore be determined easily by inspection. To determine the equivalence point without plotting a titration curve, an indicator can be used.

$$\mathbf{H^+ + B \rightarrow HB^+}$$
$$\mathbf{proton\left(from\ acid\right) + base \rightarrow base\ with\ proton}$$

Equation 6. Generic Acid-Base Reaction

$$\mathbf{pH = -\log\left[H^+\right],\ where\ [\ \ ] = concentration}$$

Equation 7. Definition of pH

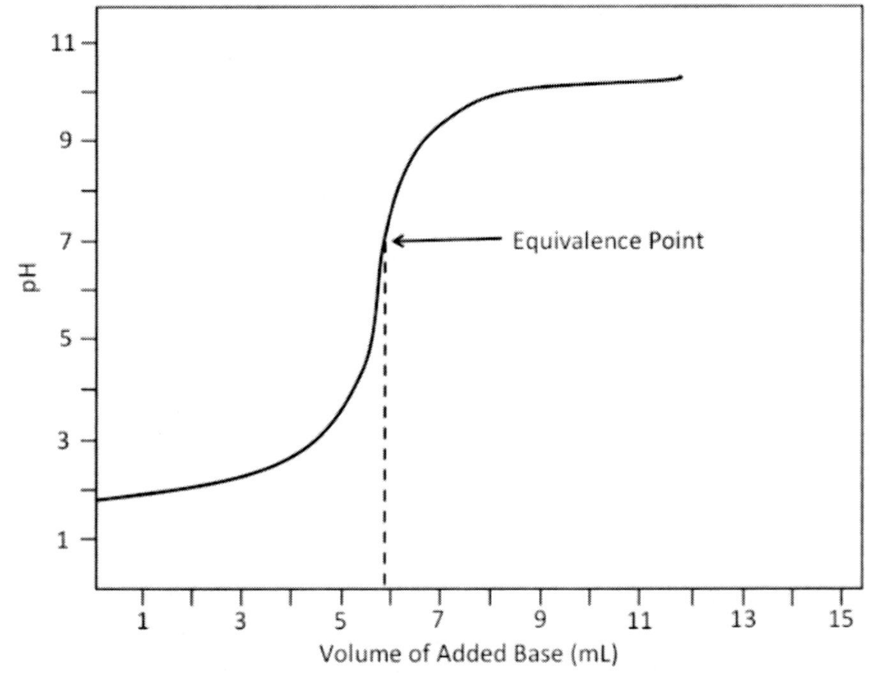

Figure 4. Sample Titration Curve

An indicator is a chemical compound that changes color as the solution containing the indicator changes pH. Different indicators have different pH ranges where they change color. The choice of indicator must be made so that the color change occurs near the pH of the equivalence point. In this experiment, you will use phenolphthalein as an indicator. Phenolphthalein is clear at acidic/neutral pH values but turns pink as the solution becomes basic. You will see this transition from clear to pink when you've reached the equivalence point of your titration. When you see this transition, determine the volume of added solution it took to reach the equivalence point and perform a calculation similar to that shown in **Example 3** to calculate the concentration of the titrated solution.

Procedure (All calculations, volumes, masses, etc. should be included in your lab report).

Part A – Preparing a 0.50 *M* NaOH Solution

1. Calculate the mass of NaOH needed to prepare 100.0 mL of a 0.50 *M* NaOH aqueous solution. Verify your calculation with your instructor before proceeding.
2. Weigh out the mass of NaOH calculated in step 1. Record the exact mass in your lab report.
3. Obtain a clean 100-mL volumetric flask and add ~50 mL of deionized or distilled water before adding the NaOH weighed in step 2. Swirl to mix the contents of the flask. Be careful as the flask may get hot during this process.
4. Add ~20 mL of additional deionized or distilled water to the flask and swirl to mix thoroughly. Make sure all of the NaOH dissolves.
5. Add more deionized or distilled water until the meniscus reaches the line on the neck of the volumetric flask. Cap the flask and then swirl and invert the flask to thoroughly mix the contents.
6. Transfer the solution to a clean 250-mL beaker. Label the beaker appropriately.

Part B – Preparing a 0.10 *M* NaOH Solution by Dilution

7. Calculate the volume of the 0.50 *M* NaOH solution necessary to make 100.00 mL of a 0.10 *M* NaOH solution. Verify your calculation with your instructor before proceeding.
8. Add ~50 mL of deionized or distilled water to a clean 100.0-mL volumetric flask.
9. Measure out the volume of your 0.50 *M* NaOH solution that you calculated in step 7. Transfer this volume to the same volumetric flask used in the previous step, swirl to mix thoroughly, and add water (with intermittent swirling) until the bottom of the meniscus reaches the line. Cap the flask and then swirl and invert to thoroughly mix the contents.
10. Transfer the solution to a clean 250-mL beaker. Label the beaker appropriately.

Part C – Standardizing the 0.50 *M* NaOH Solution

11. As stated earlier, a titration involves slowly adding one reagent to another until the equivalence point is reached. The moles of one of the reagents must be known very accurately to calculate the moles of the other. We want to determine the moles of NaOH per volume of solution (to verify the accuracy of

our solution making). Therefore, we will react our NaOH, a base, with a known number of moles of an acid, potassium hydrogen phthalate, KHP.

12. Weigh out approximately 1.5 grams of KHP into a 125-mL Erlenmeyer flask. Record the exact mass of KHP used in your lab report.
13. The molar mass of KHP is 204.22 g/mol. Calculate the number of moles of KHP based upon the mass of KHP from step 12.
14. Add roughly 30 mL of deionized or distilled water to the flask and swirl until all of the KHP dissolves.
15. Add a few drops of phenolphthalein indicator to the contents of the 125-mL flask. Remember, phenolphthalein will turn pink as the solution becomes basic.
16. To add the NaOH solution to the freshly made KHP solution in a controlled manner, you will use a piece of equipment called a buret (50-mL), shown in **Figure 5**.

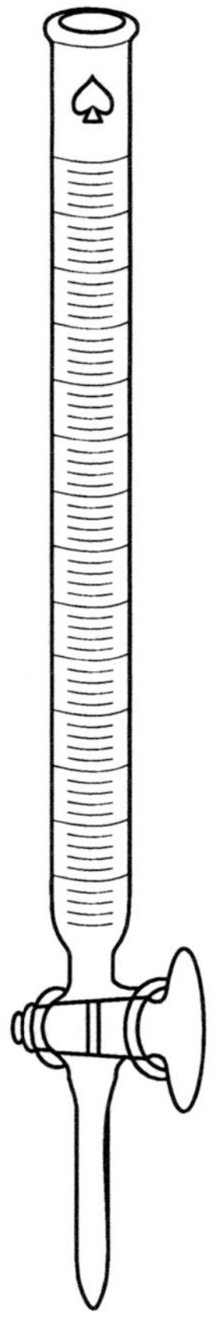

Figure 5. Buret

17. Before using a buret, it should be properly cleaned at least twice with deionized or distilled water and then at least twice with the solution to be dispensed.
18. To clean your 50-mL buret, turn the knob near the tip (called a stopcock) so that it is perpendicular to the barrel of the buret. This is the closed position. Add about 5 mL of the appropriate cleaning agent to the buret. Turn the buret on its side with the open end over a waste beaker, and rotate the barrel so that the liquid contacts the entire inside surface of the buret as it is poured out. Save one

or two mL to dispense through the tip of the buret as well. Repeat this procedure until the minimum of four cleanings listed above have been done.

19. Once your buret is clean, secure it to a ring stand using a buret clamp as shown in **Figure 6**. Close the stopcock and fill the buret to near the top with 0.50 *M* NaOH solution.

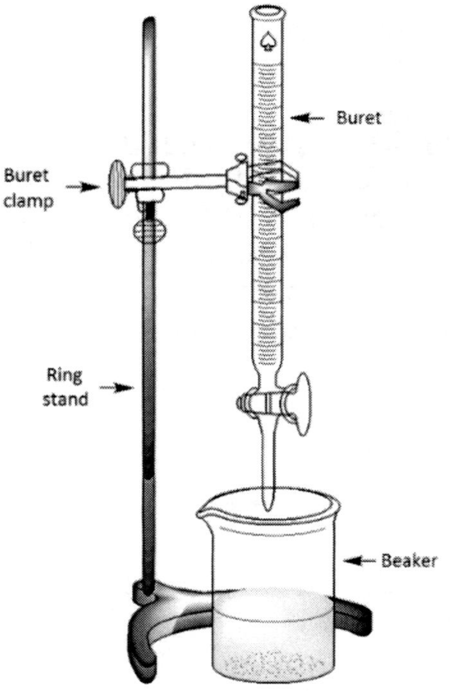

Figure 6. Buret Set-Up

20. If the solution in the buret is above the graduation marks, drain some solution into a waste beaker until the solution level is below the start of the graduation marks. Record the initial volume of the NaOH solution in your lab report. Look to see if there is a bubble in the burer tip. If present, open the stopcock fully until the bubble and some of the solution flows into a waste beaker.
21. Dispense the NaOH solution into the KHP/phenolphthalein solution in the Erlenmeyer flask while swirling the flask to ensure complete mixing. The addition can be done quickly until the first hint of pink color is seen in the flask. However, the rate of addition from the buret should still not exceed a quick flow of *drops*.
22. After the pink color is observed, even in the smallest amount, you must slow your addition immediately. Turn the stopcock to the off position.
23. Small amounts of NaOH solution can then be dispensed from the buret by quickly turning the stopcock 180°. After each turn, the flask should be swirled.
24. USE CAUTION! It is very easy to overshoot the equivalence point in a titration, requiring that you start the titration all over again. When a faint pink color lasts for 30 seconds before disappearing, you have reached the titration equivalence point. Record the final volume of the NaOH solution in the buret in your lab report.

25. Determine the volume of NaOH solution needed to reach the equivalence point.
26. Keeping in mind that 1 mole of KHP reacts with 1 mole of NaOH, and using the moles of KHP in the beaker calculated earlier, determine the moles of NaOH used to reach the equivalence point.
27. Using the moles of NaOH and the volume used to reach the equivalence point, calculate the molarity of the NaOH solution.
28. Repeat steps 12-15 and 20-27 to perform a second trial of this standardization.

Part D – Standardizing the 0.10 *M* NaOH Solution

29. Repeat steps 11-28 to perform two titrations with your 0.10 *M* NaOH solution. You will need to use 0.3 grams of KHP instead of 1.5 grams and you will use your 0.10 *M* NaOH solution in place of your 0.50 *M* NaOH solution.
30. After you are sure you have all of the data you need, dispose of any remaining solutions in the appropriate waste container and rinse all glassware thoroughly.

Experiment 12 Pre-Lab (Show ALL Work and ALWAYS include units)

1. You are asked to make 250.0 mL of a 0.56 *M* aqueous solution of NaCl. How many grams of NaCl are required to make this solution?

2. You are then asked to make 250.0 mL of a 0.040 *M* aqueous NaCl solution by diluting the appropriate amount of your previously prepared 0.56 *M* solution. What volume of the 0.56 *M* solution is required to prepare the 0.040 *M* solution?

3. A beaker containing a solution with 0.0120 moles of acid is titrated with a NaOH solution of unknown concentration from a buret. 11.3 mL of NaOH solution is required to reach the equivalence point. The reaction stoichiometry is 1 mole of NaOH reacts with 1 mole of acid. How many moles of base are needed to reach the equivalence point?

4. What is the molarity of the NaOH solution used in the previous question?

Experiment 12 Lab Report (Show ALL Work and ALWAYS include units)

Part A – Preparing a 0.50 *M* NaOH Solution

1. Calculate the mass of NaOH needed to make 100.0 mL of an aqueous solution that is 0.50 *M* in NaOH. Show all work.

2. Exact mass of NaOH used: ________ g

Part B – Preparing a 0.10 *M* NaOH Solution by Dilution

3. Calculate the volume of 0.50 *M* NaOH solution necessary to make 100.0 mL of a 0.10 *M* NaOH solution. Show all work.

Part C – Standardizing the 0.50 *M* NaOH Solution

Trial 1

4. Exact mass of KHP used: ________ g

5. Moles of KHP used (show work): ________ moles

6. Initial volume in buret: ________ mL

7. Final volume in buret: ________ mL

8. Total volume used to reach equivalence point: ________ mL

9. Moles of NaOH used (1 mol NaOH/1 mol KHP): ________ moles

10. Molarity of NaOH solution (show work): ________ *M*

Trial 2

11. Exact mass of KHP used: ________ g

12. Moles of KHP used (show work): ________ moles

13. Initial volume in buret: ________ mL

14. Final volume in buret: ________ mL

15. Total volume used to reach equivalence point: ________ mL

16. Moles of NaOH used (1 mol NaOH/1 mol KHP): ________ moles

17. Molarity of NaOH solution (show work): ________ *M*

18. Average molarity from Trial 1 and 2: ________ *M*

Part D – Standardizing the 0.10 *M* NaOH Solution

Trial 1

19. Exact mass of KHP used: ________ g

20. Moles of KHP used (show work): ________ moles

21. Initial volume in buret: ________ mL

22. Final volume in buret: ________ mL

23. Total volume used to reach equivalence point: ________ mL

24. Moles of NaOH used (1 mol NaOH/1 mol KHP): ________ moles

25. Molarity of NaOH solution (show work): ________ *M*

Trial 2

26. Exact mass of KHP used: ________ g

27. Moles of KHP used (show work): ________ moles

28. Initial volume in buret: ________ mL

29. Final volume in buret: ________ mL

30. Total volume used to reach equivalence point: ________ mL

31. Moles of NaOH used (1 mol NaOH/1 mol KHP): ________ moles

32. Molarity of NaOH solution (show work): ________ *M*

33. Average molarity from Trial 1 and 2: ________ *M*

34. Compare the average concentrations you determined by titration (standardization) with those that you originally intended when you prepared the solutions.

35. Based on your results from question 34, explain why chemists often need to standardize solutions they've made if an accurate solution concentration is important.

Experiment 13
Getting Specific with Specific Heat

Objectives
- Explore the behavior of energy in the form of heat.
- Become proficient with calorimetry.
- Think about how specific heat relates to physical states of matter.

Topics Introduced in this Lab
- Heat
- Specific heat
- Calorimetry

Related Zumdahl Sections
- Chapter 7 Sections 1 and 3

Safety Considerations
- Safety goggles and appropriate lab attire must be worn at all times when working in the lab.
- Do not ingest any of the chemicals in this experiment.
- Lead, dust in particular, is toxic. Do not ingest lead pellets, hold in a bare hand, or rub together to create atmospheric dust.

Introduction
Specific heat is an intensive property of a substance that is defined as the amount of energy required to raise one gram of a substance by one degree Celsius. Water has a specific heat (C) of $4.18\frac{J}{g°C}$. This value means that it would take 4.18 joules of energy to raise 1 gram of water by 1 degree Celsius. Similarly, 4.18 joules of energy would have to be lost from 1 gram of water to lower its temperature by 1 degree Celsius.

In this experiment, you will determine the specific heats of two different metals, lead and copper. You will then discuss your findings. To accomplish these tasks, it is important to understand two key points about energy in the form of heat.

1) Energy in the form of heat will flow spontaneously from a substance having a higher temperature to a substance having a lower temperature.
2) Energy does not simply disappear. It may be transferred from one substance to another, but the total amount of the energy is conserved.

Energy in the form of heat (q) that is transferred into or out of a substance is related to the specific heat (C) of that substance according to **Equation 1**.

$$q = mC\Delta T$$

Equation 1. Heat Transferral Equation

From this, we can see that if we know the heat that is transferred into or out of a substance (in joules), the substance's mass (in grams), and its change in temperature (in °C), we can calculate its specific heat in J/g°C. Determining the mass of a substance and its temperature change are easy enough using a balance and a thermometer, but how do we measure the heat gained or lost by a substance? The answer is calorimetry! Calorimetry is a method to measure the transfer of energy in the form of heat.

To understand how calorimetry works, imagine heating a cube of metal until it is red hot. Since heat flows spontaneously from higher temperature substances to lower temperature substances and since energy must be conserved, we can make the statement that the heat that is lost by the metal cube as it cools is not truly lost but is instead simply transferred to the cube's surroundings. Any energy in the form of heat that is lost by the metal MUST be gained by the surroundings. Mathematically, we can state this as **Equation 2.**

$$q_{\text{metal}} + q_{\text{surroundings}} = 0$$

or

$$-q_{\text{metal}} = +\,q_{\text{surroundings}}$$

Equation 2. Heat Loss and Gain Equation

The "–" sign is a convention used to indicate that heat is flowing out of the metal and the "+" sign is used to indicate that heat is flowing into the surroundings. Combining **Equations 1** and **2** the expression becomes **Equation 3**.

$$-[(m_{\text{metal}})(C_{\text{metal}})(\Delta T_{\text{metal}})] = [(m_{\text{surroundings}})(C_{\text{surroundings}})(\Delta T_{\text{surroundings}})]$$

Equation 3. Expanded Heat Loss and Gain Equation

Trying to measure the mass, specific heat, and ΔT of all surroundings is impossible. Therefore, if we isolate a small system, the heat given off by the metal cube can be easily measured and all of the relevant values determined. To do that, we place the metal cube into water, which is in turn inside an insulating container called a calorimeter. The water and the calorimeter are the surroundings. The insulating container prevents heat from escaping, effectively isolating the system from other surroundings. In this system, the metal cube releases heat while both the water AND the calorimeter gain that energy in the form of heat. In other words,

$$-q_{\text{metal}} = +q_{\text{water}} + q_{\text{cal}}$$

Equation 4. Heat Loss and Gain Equation for Calorimeter System

Keeping in mind that $q = mC\Delta T$, the above expression expands to become **Equation 5**.

$$-[(m_{\text{metal}})(C_{\text{metal}})(\Delta T_{\text{metal}})] = [(m_{\text{water}})(C_{\text{water}})(\Delta T_{\text{water}}) + (m_{\text{cal}})(C_{\text{cal}})(\Delta T_{\text{cal}})]$$

Equation 5. Expanded Heat Loss and Gain Equation for Calorimeter System

Since a single calorimeter can be made of multiple materials, determining its specific heat (C_{cal}) can be problematic. Instead, we can combine the mass and specific heat of the calorimeter into a single term X_{cal} (in J/°C), which can be experimentally determined and then used in subsequent experiments *as long as the same calorimeter is used for all of the experiments.* Therefore, our expression for heat loss/gain becomes **Equation 6.**

$$-[(m_{metal})(C_{metal})(\Delta T_{metal})] = [(m_{water})(C_{water})(\Delta T_{water}) + (X_{cal})(\Delta T_{cal})]$$

Equation 6. Final Expression for Heat Loss and Gain

At this point, all values in the above expression can be measured or are known except for the specific heat of our metal cube (C_{metal}) and X_{cal}. Therefore, once we know the value of X_{cal} the specific heat of the metal can be determined.

To experimentally determine X_{cal} requires no new concepts. We simply need to substitute our hot metal with something else hot that has a known specific heat. Water fills this role nicely! Using hot water instead of hot metal changes our expression slightly

$$-[(m_{hot\ water})(C_{hot\ water})(\Delta T_{hot\ water})] = [(m_{water})(C_{water})(\Delta T_{water}) + (X_{cal})(\Delta T_{cal})]$$

Equation 7. Heat Loss and Gain Equation Modified for X_{cal} Determination

The only value now that is not known or can be measured is X_{cal}. So we can solve for X_{cal}, then use it to determine specific heat values in future experiments *where we use the same calorimeter.*

Therefore, the general procedure for a calorimetry experiment is as follows:

1) Determine the value of X_{cal} for your calorimeter.
   - –Do this by mixing known masses of cool and hot water in your calorimeter and measuring the temperature change that occurs for both. (The temperature change for the cool water will also be the temperature change for the calorimeter.)
   - –Use Equation 7 to calculate X_{cal}.
2) Determine the specific heat (C_{metal}) of your substance of interest.
   - –Mix a known mass of the heated substance with a known mass of cool water in the same calorimeter you used to determine X_{cal}.
   - –Measure the temperature change for both the hot substance and cool water. (The temperature change for the cool water will also be the temperature change for the calorimeter.)
   - –Use Equation 6 to calculate C_{metal}.

You will use this general procedure in this experiment to determine the specific heat of solid lead and copper.

Procedure (*Be sure to record all relevant data in your lab report.)

Part A – Determining X_{cal} for Your Calorimeter

1. Obtain two styrofoam cups and a lid. Put one cup inside the other. This set of cups, along with the lid, will serve as your calorimeter. Add a stir bar to your calorimeter and place it on a stir plate.
2. Slide a thermometer through the hole in the lid. This thermometer will be used to measure the temperature of the contents in the calorimeter. Use a slit stopper, ring stand, and a utility clamp to support the thermometer in the calorimeter as shown in **Figure 1**. Ensure that the stir bar will not hit the thermometer.
3. Add 25.0 mL of cool water (room temperature) to your calorimeter and turn on the stir plate to begin stirring. Record the exact volume of cool water used.
4. Heat 70.0 mL of water to approximately 80 °C in a beaker.
5. After letting the cool water sit in the calorimeter for some time, record the temperature of the cool water.

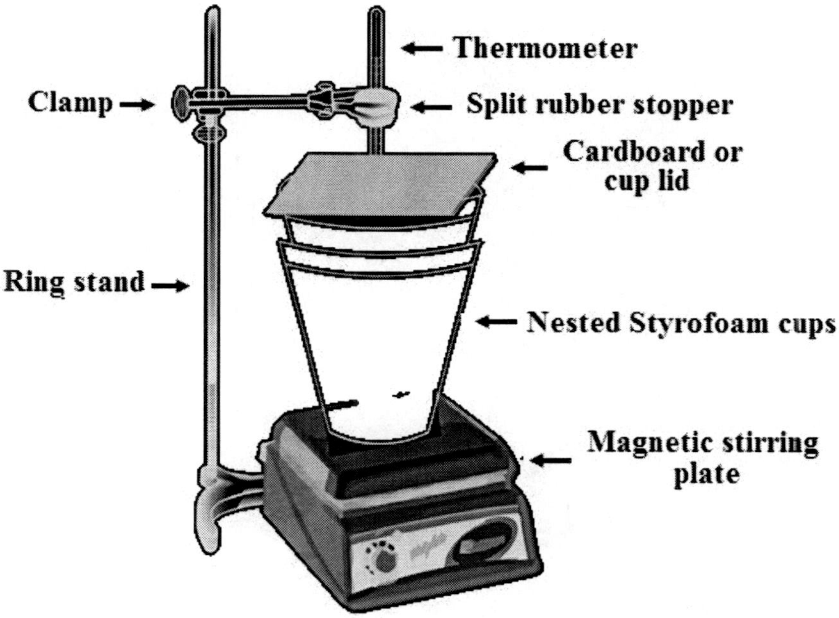

Figure 1. Calorimeter Set-up

6. Carefully measure out 25.0 mL of the hot water in a graduated cylinder. Record the exact volume used.

7. Using a second thermometer, record the temperature of the hot water in the graduated cylinder and *immediately* pour the hot water into the calorimeter containing your cool water.
8. Closely monitor the temperature of the water in the calorimeter and record the value when it equilibrates. This value is the final temperature of your system.
9. Dump the water from the calorimeter down the drain, dry your calorimeter thoroughly, and repeat steps 1–8 to perform a second trial.

<u>Part B – Determining the Specific Heat of Lead</u>

10. Set up your clean and dry calorimeter as you did in steps 1–3 in Part A, again with 25.0 mL of room temperature water.
11. Place 40 grams of lead shot into a large test tube. Record the exact mass of lead used.
12. Tilt the test tube on its side to spread the lead shot along the wall of the test tube. Insert a dry thermometer into the test tube and center it. Tilt the test tube right side up so that the lead shot fills in around the thermometer. The thermometer should not touch the walls or the bottom of the test tube (see **Figure 2**).

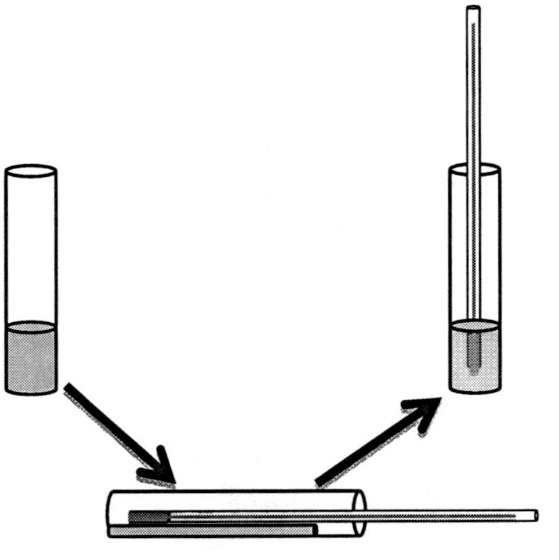

Figure 2. How to Insert the Thermometer

13. Fill a 400- or 600-mL beaker 2/3 full with water. Heat the water to boiling on a hot plate.
14. Secure the test tube with a utility clamp and ring stand so that the portion of the test tube containing the lead is submerged in the water bath (see **Figure 3**). Continue to heat the metal in this manner until the temperature no longer increases.
15. Record the temperature of the heated metal and the temperature of the room temperature water in your calorimeter.

16. Pour the heated lead shot into your calorimeter. Closely monitor the temperature of the contents in the calorimeter and record the value when it stops changing. This value is the final temperature of your system.
17. Dispose of the contents of your calorimeter according to the instructions given to you by your instructor.
18. Repeat steps 10-17 to perform a second trial to determine the specific heat of lead.

Figure 3. Heating the Metal

<u>Part C – Determining the Specific Heat of Copper</u>

19. Repeat Part B using 40 grams of copper shot instead of lead for each trial (2 trials should be performed). Be sure to record the exact mass of copper used in each trial.

Experiment 13 Pre-Lab (Show ALL Work and ALWAYS include units)

1. Is it important to know the mass of the calorimeter that you use? Why or why not?

2. To determine X_{cal} for a calorimeter, you add 25.6 mL of water at 75.3 °C to 24.8 mL of water at 20.3 °C that has been sitting in a calorimeter. While stirring, you monitor the temperature of the mixture until it no longer changes, and measure the final temperature of the system to be 44.6 °C. Based upon this data, calculate X_{cal} for this calorimeter.

3. You add 40.343 g of a metal at 93.4 °C to a calorimeter containing 25.1 mL of water at 20.3 °C. After letting the system equilibrate, you measure the temperature of the system to be 23.4 °C. Assuming that you used the same calorimeter as in the previous question, calculate the specific heat of the metal.

Experiment 13 Lab Report (Show ALL Work and ALWAYS include units)

Part A – Determining X_{cal} for Your Calorimeter

1. Volume of cool water used: Trial 1: ________ Trial 2: ________

2. Temperature of cool water: Trial 1: ________ Trial 2: ________

3. Volume of hot water used: Trial 1: ________ Trial 2: ________

4. Temperature of hot water: Trial 1: ________ Trial 2: ________

5. Final temperature of system: Trial 1: ________ Trial 2: ________

6. Calculation of X_{cal} (Trial 1):

7. Calculation of X_{cal} (Trial 2):

8. Average X_{cal} from Trials 1 and 2: ________

Part B – Determining the Specific Heat of Lead

9. Volume of cool water used: Trial 1: ________ Trial 2: ________

10. Temperature of cool water: Trial 1: ________ Trial 2: ________

11. Mass of metal used: Trial 1: ________ Trial 2: ________

12. Temperature of hot metal: Trial 1: ________ Trial 2: ________

13. Final temperature of system: Trial 1: ________ Trial 2: ________

14. Calculation of C_{lead} (Trial 1):

15. Calculation of C_{lead} (Trial 2):

16. Average C_{lead} from Trials 1 and 2: ________

Part C – Determining the Specific Heat of Copper

17. Volume of cool water used: Trial 1: __________ Trial 2: __________

18. Temperature of cool water: Trial 1: __________ Trial 2: __________

19. Mass of metal used: Trial 1: __________ Trial 2: __________

20. Temperature of hot metal: Trial 1: __________ Trial 2: __________

21. Final temperature of system: Trial 1: __________ Trial 2: __________

22. Calculation of C_{copper} (Trial 1):

23. Calculation of C_{copper} (Trial 2):

24. Average C_{copper} from Trials 1 and 2: ________

25. Look up the densities for lead, copper, and water and record those values here (be sure to cite your source(s)).

<u>Substance</u> <u>Density (g/mL)</u>

Lead

Copper

Water

Source(s):

26. Using the densities you recorded above along with the specific heat values you experimentally determined (or know in the case of water), is there a trend of how specific heat relates to the physical state of these substances and why?

27. Comparing lead and copper, do the specific heat values you determined support your explanation? Why or why not?

Experiment 14

Mass and Density of a Gas – Determining the Mass, Density, and Moles of a Gas

Objectives

- Measure the mass of a gas generated in the lab.
- Using the measured mass, determine the density and moles of the gas.

Topics Introduced in this Lab

- Ideal gas law
- Mass of a gas
- Density of a gas

Related Zumdahl Sections

- Chapter 8 Sections 1-4

Safety Considerations

- Acetic acid, CH_3COOH, present in vinegar, should not come into contact with your eyes. Handle with care.
- $NaHCO_3$, baking soda, is slightly caustic. Handle with care.
- Do not ingest any of these compounds.
- Safety glasses and appropriate lab wear must be worn at all times when working in the lab.

Introduction

The simplest state of matter is the gas phase. In this phase, the atoms, ions, or molecules are relatively far apart from each other. You come in contact with many gases in your life. Air is a mixture of gases, primarily nitrogen and oxygen. Steam is water in the gas phase. Helium-filled balloons contain gaseous helium. Because many gases are clear and colorless, some people conclude that gases do not have mass. That is not true. In this experiment we will make a gas, weigh it, and determine its density.

Because gases are a simple state of matter, they are described with a simple equation of state called the ideal gas law, **Equation 1**. Since R is a constant, **Equation 1** indicates that if we know three of the four quantities, P, V, n, or T, we can calculate the remaining one.

$$PV = nRT$$

where P = gas pressure in atmospheres
V = gas volume in liters
n = moles of gas

R = ideal gas constant = $0.08206 \dfrac{\text{L} \cdot \text{atm}}{\text{mol} \cdot \text{K}}$
T = gas temperature in K

Equation 1. Ideal Gas Law

For example, if a helium-filled balloon contains 1.50 L of helium at 1.25 atm of pressure and at a temperature of 23.4°C (296.6 K), how many moles of He gas are in the balloon? What is the mass of the He gas in the balloon?

$$PV = nRT, \text{ thus } n = \frac{PV}{RT}$$

$$n = \frac{(1.25 \text{ atm})(1.50 \text{ L})}{(0.08206 \frac{\text{L} \cdot \text{atm}}{\text{mol} \cdot \text{K}})(296.6 \text{ K})}$$

$$n = 7.70 \times 10^{-2} \text{ mol}$$

molar mass of He is 4.003 g/mol

$$(7.70 \times 10^{-2} \text{ mol})(4.003 \text{ g/mol}) = 0.308 \text{ g}$$

In this experiment we will generate carbon dioxide, $CO_2(g)$, from the chemical reaction of acetic acid with sodium hydrogen carbonate, **Equation 2**. Vinegar is a good cheap source of acetic acid. Baking soda is nearly pure sodium hydrogen carbonate. Then we will measure the mass of $CO_2(g)$ captured in a soda pop bottle. Using the mass of $CO_2(g)$ collected and the volume of the soda pop bottle we can determine density of the collected $CO_2(g)$.

$$\mathbf{CH_3COOH(\mathit{aq}) + NaHCO_3(\mathit{s}) \rightarrow CO_2(\mathit{g}) + NaCH_3COO(\mathit{aq}) + H_2O(\mathit{l})}$$

Equation 2. Reaction of Vinegar and Baking Soda

Procedure

Part A – Generate CO_2 Gas

1. Obtain a 250-mL filter flask, one-hole rubber stopper, 50-mL buret and plastic or rubber tubing. Also get a clean and dry 500-mL soda pop bottle complete with screw cap.
2. Using an analytical balance, determine the mass of the empty soda pop bottle including the screw cap.
3. Secure the soda pop bottle using a ring stand and either a utility clamp or iron ring. Also secure the 250-mL filter flask using a different ring stand and utility clamp.
4. Attach one end of the plastic or rubber tubing to the filter flask nipple then insert the other end into the open soda pop bottle. The plastic or rubber tubing must not completely fill the mouth of the soda pop bottle or a highly pressurized (and potentially dangerous) system could result.
5. Place approximately 10.0 g of $NaHCO_3$ (baking soda) into the bottom of the filter flask. Firmly insert the one-hole rubber stopper into the 250-mL filter flask top.

6. Attach the 50-mL buret to the same ring stand as the 250-mL filter flask using a buret or utility clamp. Align the buret tip with the hole in the rubber stopper then lower the buret until its tip is barely inserted into the rubber stopper hole.
7. Pour approximately 50.0 mL of vinegar into the closed buret.
8. Open the buret stopcock so that a small steady stream of vinegar drops into the filter flask. **Go slowly as you do not want the liquid mixture created as the reaction proceeds to rise enough that it flows into the tubing**. As the gas is created you will have to increase the vinegar flow from the buret to counteract the pressure increase.
9. After the buret has emptied close the stopcock, refill it with an additional 50.0 mL then slowly add the additional 50.0 mL of vinegar to the filter flask.

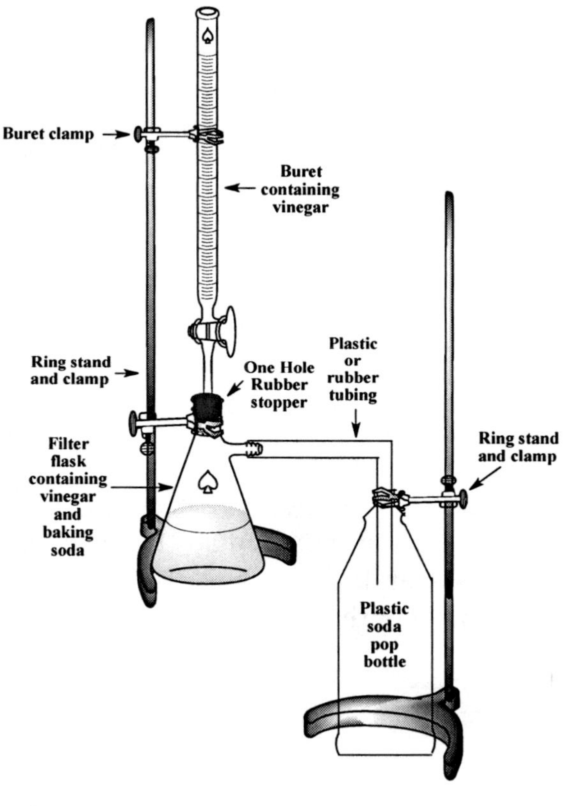

Figure 1. Experimental Setup

Part B – Determine Mass of $CO_2(g)$ and Volume of the Bottle

10. Remove the tubing from the soda pop bottle then screw the bottle cap onto the soda pop bottle.
11. Using the same analytical balance as before, determine the mass of the filled soda pop bottle including the screw cap. This will be the mass of the bottle plus the mass of the collected $CO_2(g)$.
12. Open the soda pop bottle and fill it with water.
13. Measure the water volume by pouring it into a graduated cylinder.
14. Record the temperature of the room in which you are working in your lab report.
15. Get the current atmospheric pressure from your instructor and record that value in your lab report.
16. Repeat the entire procedure for a second trial.

Experiment 14 Pre-Lab

1. Vinegar is a 5.0% solution of acetic acid, CH_3COOH, dissolved in water. (There are 5.0 g of acetic acid for every 100.0 g of vinegar.) Vinegar has a density of 1.01 g/mL. How many mL of vinegar are required to completely react with 6.25 g of $NaHCO_3$?

2. How many grams of $CO_2(g)$ will be generated from the reaction described in question 1?

3. A clean and dry 375-mL soda pop bottle with screw cap has a mass of 29.85 g. When filled with $CO_2(g)$, the same soda pop bottle and screw cap has a mass of 30.97 g. When filled with water, the soda pop bottle is found to have a volume of 502.4 mL. What is the density of the CO_2 gas in this experiment?

4. Using the density determined in question 3, calculate the mass of $CO_2(g)$ required to completely fill a 2.0-L soda pop bottle.

Experiment 14 Lab Report

Part A – Generate CO_2 Gas

	Trial 1	Trial 2
1. Mass of empty soda pop bottle including screw cap	______	______

Part B – Determine Mass of $CO_2(g)$ and Volume of the Bottle

	Trial 1	Trial 2
2. Mass of soda pop bottle including screw cap when filled with $CO_2(g)$	______	______
3. Volume of soda pop bottle	______	______
4. Room temperature	______	______
5. Atmospheric pressure	______	______

6. Using your measured values, determine the density of $CO_2(g)$ for trials 1 and 2.

7. Calculate the number of moles and molecules of $CO_2(g)$ contained in the 500-mL soda pop bottle for both trials.

8. Why can we use atmospheric pressure as the pressure of the gas in the soda pop bottle?

9. When you originally weighed the empty bottle and screw cap, the bottle contained air, a gas. However, the balance did not indicate a mass for the air. When the bottle filled with CO_2 was weighed, you observed a mass for the gas. Explain.

10. If the soda pop bottle initially was weighed dry but was not completely dry after the $CO_2(g)$ was delivered into the bottle, how would that affect the determined density of $CO_2(g)$? Explain your answer.

11. Exhaled air from humans typically is 16.0% O_2, 78.1% N_2, 4.0% CO_2 by volume, and the rest is mostly noble gases. In a normal exhaled breath adult humans exhale 500.0 mL. Using your determined $CO_2(g)$ density, calculate the number of CO_2 molecules that humans exhale in every typical breath.

Experiment 15
Melting Point Differences –
Effect of Intermolecular Forces On Melting Points

Objective
- See the effects of intermolecular force differences on melting points.

Topics Introduced in this Lab
- Intermolecular forces

Related Zumdahl Sections
- Chapter 9 Sections 2, 6, and 7
- Chapter 10 Sections 1 and 5

Safety Considerations
- Molten wax will ignite if too near a flame. Be careful when melting the wax.
- Safety goggles and appropriate lab attire must be worn at all times when working in the lab.

Introduction
Chemical compounds have attractive forces both within an individual particle of the substance and between particles. Intramolecular forces occur within the substance particles. Covalent bonds are good examples of intramolecular forces. Intramolecular force strength is equivalent to bond energies. Forces between particles, *intermolecular* forces, determine the physical states of substances. Whether or not a substance is a solid, liquid, or gas at room temperature and the temperature at which it melts or boils at a constant pressure are determined by intermolecular forces. **Figure 1** illustrates the distinction between intramolecular and intermolecular forces for carbon dioxide.

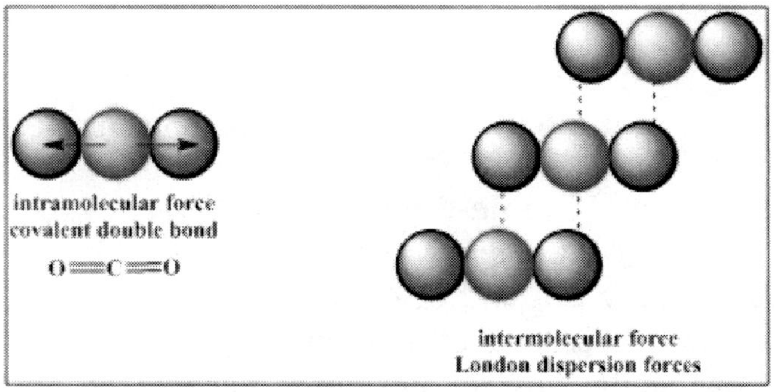

Figure 1. Representation of Intramolecular and Intermolecular Forces in Carbon Dioxide

Intramolecular forces, chemical bonds, are much stronger than intermolecular forces. Typically, chemical bonds are 100 to more than 1000 times stronger than intermolecular forces. When a molecular substance melts or boils, the molecules are not being torn apart into their constituent atoms. They are simply separating themselves from one another, which decreases the intermolecular attraction to their neighbors. For example, when water boils, it does not form hydrogen atoms/molecules and oxygen atoms/molecules, but it remains independent water molecules that are much farther from one another than in the liquid form. In other words, boiling water changes liquid water to gaseous water.

Intermolecular forces are classified in two basic types that depend upon whether or not the interacting substances are polar or nonpolar. Polar molecules, like PH_3 or H_2O, are attracted to each other via dipole-dipole forces. Dipole-dipole forces are due to the attraction of the partially negative part of a molecule to the partially positive part of an adjacent molecule as illustrated in **Figure 2**. Because oxygen atoms are more electronegative than phosphorus atoms, the partially negative areas of the H_2O molecules are more negative than is the case for PH_3. The net result is that the dipole-dipole force in H_2O is significantly larger than in PH_3. Chemists make a distinction between the weaker and stronger dipole-dipole forces, calling the attraction in water hydrogen bonding, a very strong version of dipole-dipole forces. To form a hydrogen bond the central atom in the molecule must be a N, O, or F atom, the three most electronegative elements.

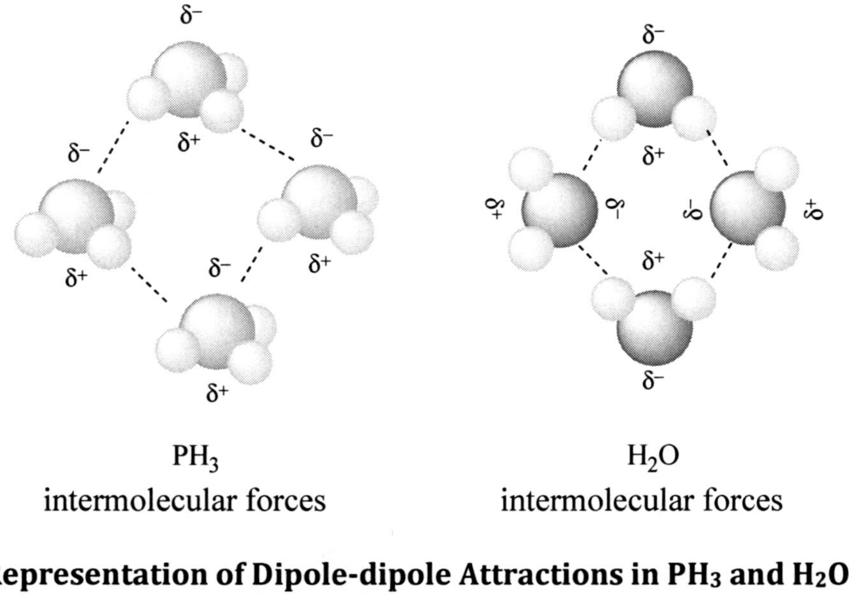

Figure 2. Representation of Dipole-dipole Attractions in PH_3 and H_2O

Nonpolar substances are found in nature as solids, liquids, and gases. Thus there are intermolecular forces, called London dispersion forces, present in these substances. Nonpolar molecules do not have permanent partially positive or negative molecular parts. However, because the electrons in these molecules are constantly in motion, partially positive and negative parts of the molecule are generated that exist only for nano- or microseconds. These fleeting areas of partial charge are sufficient to attract molecules in the liquid or solid state (see **Figure 3**). London dispersion forces (named after Fritz London who proposed their existence) are the result of these instantaneously generated partially positive and negative molecular areas. London dispersion forces are the weakest intermolecular forces.

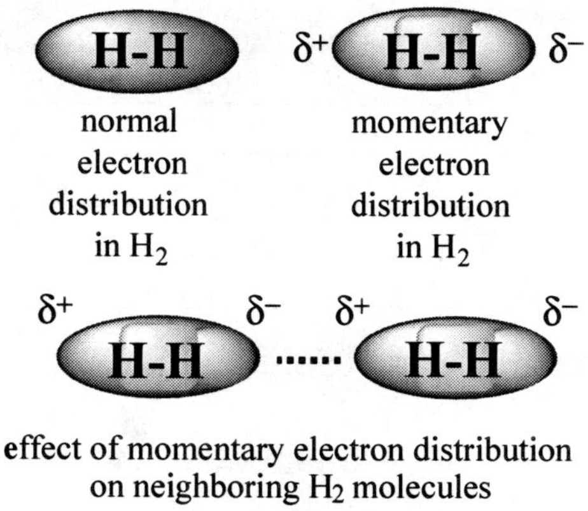

Figure 3. Representation of London Dispersion Forces in H_2

Solids are classified based upon the components that occupy lattice points in the solid structure. Those components also influence the attractive forces inside the solid. **Table 1** is a compilation of the common solid types.

Solid Type	Component Occupying Lattice Point	Attractive Force	Examples
Metallic	Metal atom	Nondirectional covalent bond	Au, Na, K, Pb
Network	Nonmetal atom	Directional covalent bond	graphite, diamond, carborundum
8A	Group 8A atom	London dispersion	He, Ne, Ar, Kr, Xe, Rn
Molecular	Molecule	Dipole-dipole, hydrogen bond, London dispersion	AsH_3, NH_3, I_2
Ionic	Ions	Ionic bond	KI, $MgCl_2$, AlI_3

Table 1. Classification Scheme for Solids

Melting and boiling temperatures at 1.00 atmosphere of pressure, frequently called normal melting and boiling points, are good indicators of intermolecular force strength. Strong intermolecular forces result in high melting and boiling points as shown in **Table 2**. **Table 2** indicates network solids have the highest melting and boiling points by far. Ionic solids are next highest, followed by metallic solids. Lowest melting and boiling points generally occur in 8A and molecular solids. While these are good trends to know, there are some exceptions. In particular, metallic solids have enormous variations in melting and boiling points. For example, Hg melts at −38.3 °C while W melts at 3410 °C.

Solid Type	Example	Melting Point (°C)	Boiling Point (°C)
8A	Ar	−189.4	−185.9
Molecular	PH_3	−133	-87.7
Metallic	Na	97.7	883
Ionic	$MgCl_2$	714	1412
Network	Graphite	~3652−3697	~4250

Table 2. Melting and Boiling Points of Selected Solids

In this experiment, we will examine the melting temperatures of some representative molecular, ionic, and network solids. Because some of these solids melt at temperatures greater than the typical lab thermometer can register, we must use a different indicator of heat transfer. Examining the heat transfer equation for substances in the solid phase, **Equation 1**, we see that if we keep the mass times specific heat portion of the equation constant, heat transfer is proportional to temperature change. By using the same Bunsen burner and evaporating dish as well as constant mass/specific heat combinations, the time to heat a substance to its melting point is a good relative indicator of melting-point temperature. It is important that you use the mass of each substance indicated in the procedure to equalize the heat transfer process. You must also use the same evaporating dish and Bunsen burner in your experimental setup, **Figure 4**.

$$q = mC_s\Delta T$$

where q = heat transfer

m = mass

C_s = specific heat of solid

ΔT = change in temperature

Equation 1. Heat Transfer Equation

Procedure

Part A – Determine Time Required to Melt 25.0 Grams of Wax

1. Using a balance, weigh out 25.0 g of paraffin wax.
2. Place the 25.0 g of wax into a clean and dry evaporating dish.
3. Light your Bunsen burner and adjust it to produce a distinct blue flame with an inner blue cone. Adjust the flame height so that the inner blue cone will just touch the bottom of the evaporating dish when inserted beneath the dish.
4. Assemble the experiment so that it looks like **Figure 4**, adding the Bunsen burner last. (This step can be done inside a fume hood or fume extractor to minimize smoke emission into the room.)

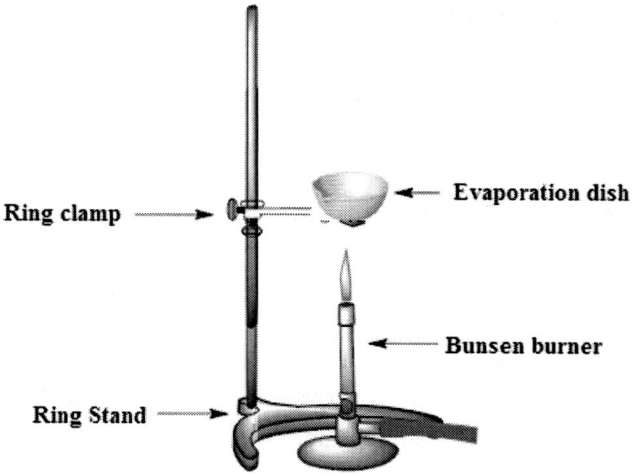

Figure 4. Experimental Setup

5. Once the Bunsen burner is in place, begin timing the experiment.
6. Record the time starting from when you placed the Bunsen burner underneath the evaporating dish until all of the wax melts. This will require careful observation.
7. After the wax has melted, turn off the Bunsen burner and allow the evaporating dish and wax to cool to room temperature. Do NOT change the settings (air intake or gas flow regulator knobs) on the Bunsen burner. Also, do NOT change the height of the ring clamp.
8. Carefully clean the evaporating dish to remove all of the wax.

<u>Part B – Determine Time Required to Melt 83.9 Grams of Table Salt</u>

9. Using a balance, weigh out 83.9 g of table salt, NaCl.
10. Pour the 83.9 g of table salt into the clean and dry evaporating dish.
11. Re-light your Bunsen burner (without changing the settings) and repeat steps 4 to 8 for the table salt melting process. (If your salt has some residual moisture it is possible that a few salt crystals will be expelled. Be careful as hot salt crystals could burn you.)
12. Record the time necessary to melt all 83.9 g of table salt. If after 30 minutes the salt has not melted you may stop.

<u>Part C – Determine Time Required to Melt 87.4 Grams of Dry Sand</u>

13. Using a balance, weigh out 87.4 g of dry sand.
14. Pour the 87.4 g of dry sand into a clean and dry evaporating dish.
15. Re-light your Bunsen burner (without changing the settings) and repeat steps 4 to 8 for the sand melting process.
16. Record the time necessary to melt all 87.4 g of sand, if it does melt. If after 30 minutes the sand has not melted you may stop.

Experiment 15 Pre-Lab (Show ALL Work and ALWAYS include units)

1. Assign each of these substances as a metallic, molecular, network, 8A, or ionic solid.

$CaSO_4$	C_6H_6 (benzene)	Pt (platinum)	Ar (argon)	graphite
________	________	________	________	________

2. Based upon the solid type, arrange the substances given in question 1 in order of their expected melting points (from lowest to highest melting point).

3. Solid substance A has a specific heat of 0.75 J/g°C. Solid substance B has a specific heat of 1.75 J/g°C. If this experiment uses 75.0 g of substance A, how many grams of B are necessary to make the heat transfer directly proportional to the change in temperature?

4. Based upon the structures shown, determine if the substance is a molecular, ionic, or network solid.

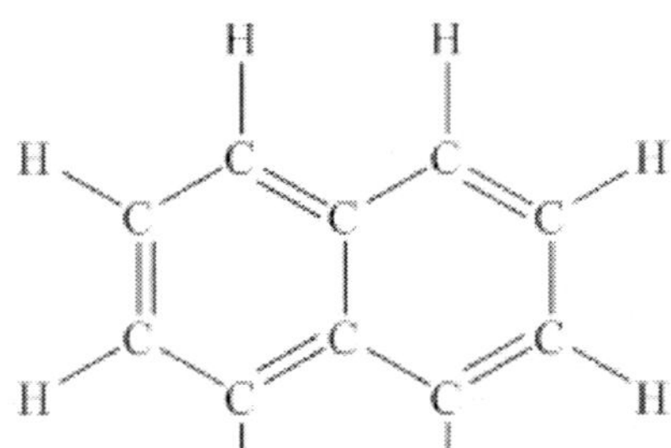

napthalene

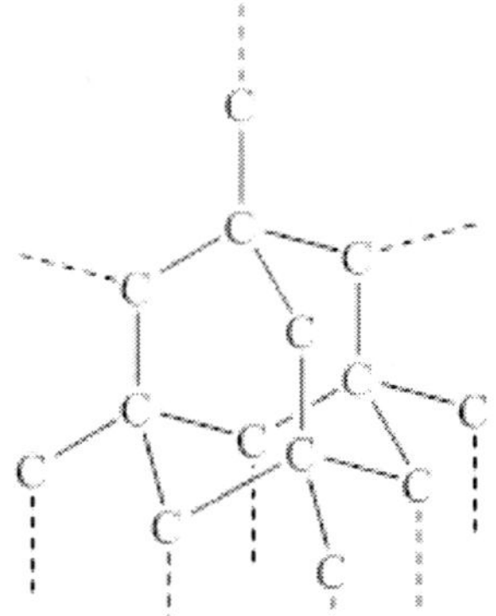

diamond

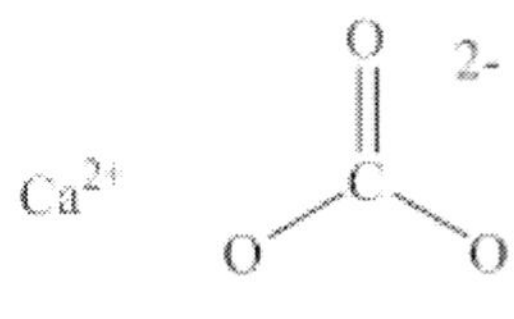

calcium carbonate

Experiment 15 Lab Report (Show ALL Work and ALWAYS include units)

1. Total time required to melt 25.0 g of wax ________

2. Total time required to melt 83.9 g of salt ________

3. Total time required to melt 87.4 g of sand ________

4. Based upon your experimental data, the substance with the highest normal melting point is ________ .

5. Based upon your experimental data, the substance with the lowest normal melting point is ________ .

6. Based upon your experimental data, is wax a molecular, ionic, or network solid? Explain your answer. Be sure to discuss what attractive force is being overcome when wax melts.

7. Based upon your experimental data, is salt a molecular, ionic, or network solid? Explain your answer. Be sure to discuss what attractive force is being overcome when salt melts.

8. Based upon your experimental data, is sand a molecular, ionic, or network solid? Explain your answer. Be sure to discuss what attractive force is being overcome when sand melts.

Experiment 16
Freezing-Point Depression – Antifreeze for Your Car

Objectives
- See how antifreeze prevents water from freezing in your car engine.
- Experience the van't Hoff factor.

Topics Introduced in this Lab
- Freezing-point depression
- Molality

Related Zumdahl Sections
- Chapter 9 Sections 2, 6, and 7
- Chapter 10 Sections 1 and 5

Safety Considerations
- $HO\text{-}CH_2\text{-}CH_2\text{-}OH$, ethylene glycol, is poisonous if ingested. It can also be absorbed through the skin. Handle with care.
- NaCl, sodium chloride, is an eye irritant.
- If any of these compounds come into contact with your eyes or skin, immediately flush the affected area with water.
- Do not ingest any of these compounds.
- Safety goggles and appropriate lab attire must be worn at all times when working in the lab.
- Ethylene glycol must be disposed of in an environmentally friendly fashion. After the lab is completed, dispose of it and any solutions containing it in a departmentally approved waste container. Do not pour ethylene glycol down the sink.

Introduction
Water has many properties that make it unique in the chemical world. One unusual property is that solid water (ice) floats in liquid water. For this to occur solid ice must be less dense than liquid water. An equivalent mass of ice has a larger volume than liquid water. Water expands as it freezes. This is why a can of soda pop placed in the freezer section of your refrigerator ruptures once the soda pop freezes. If the water in your car engine froze, it would split the engine block, destroying the engine. At that point the engine would have to be replaced at considerable cost. Clearly that is not acceptable. We add antifreeze to the car radiator to prevent water inside the engine from freezing. We also add antifreeze to the windshield washer fluid to prevent it from freezing.

When water (or any other substance) freezes, the substance particles must approach one another and interact to form a solid, **Figure 1**. If another substance is present in the solution that is freezing, particles of the second substance delay the original substance freezing by essentially blocking solid formation, **Figure 2**. For the solution to freeze, the temperature must be colder than the freezing temperature of the pure substance to mitigate the interference of the foreign substance. This is why antifreeze works in a car engine or windshield washer fluid.

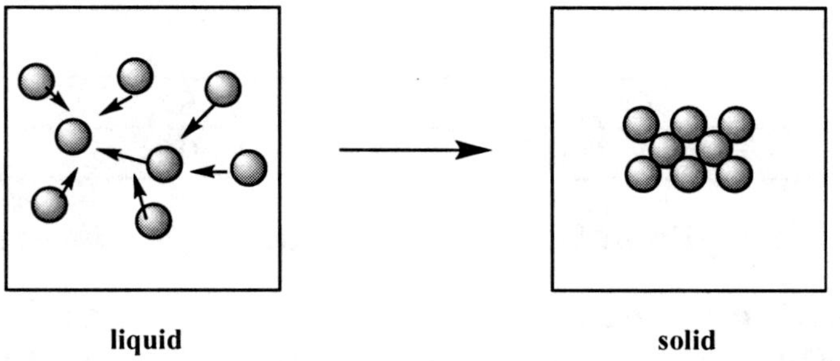

Figure 1. Atomic Representation of a Pure Substance Freezing

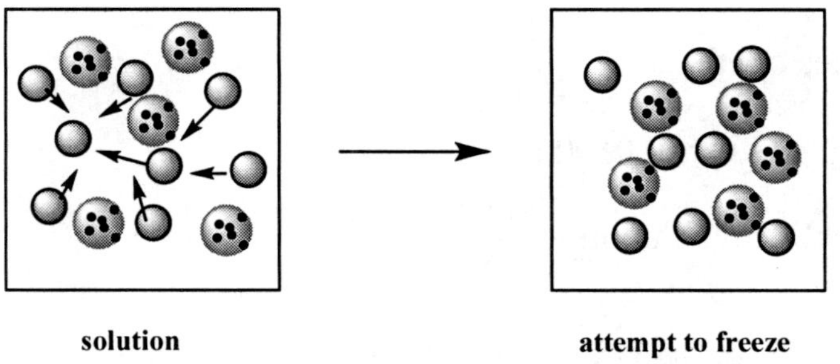

Figure 2. Atomic Representation of a Solution Freezing

The difference between the freezing temperature of a pure substance and that of a solution is the freezing-point depression. How much the freezing temperature is depressed depends on two quantities, molal concentration of the solute and a freezing-point depression constant that is unique to every solvent. **Equation 1** is the relevant mathematical relationship for freezing-point depression.

$$\Delta T_f = k_f m$$

where ΔT_f is the freezing-point depression in °C

k_f is the molar freezing-point depression constant in °C / molal

m is the molal concentration of the solution

Equation 1. Freezing-Point Depression Relationship

Molal freezing-point depression constants have been determined for many substances. Several are given in **Table 1**.

Substance	Formula	Freezing Point (°C)	k_f (°C/molal)
Water	H_2O	0.00	1.86
Acetic acid	CH_3CO_2H	16.60	3.59
Benzene	C_6H_6	5.45	5.06
Carbon tetrachloride	CCl_4	-22.8	29.8
Cyclohexane	C_6H_{12}	6.55	20.2
Ethylene glycol	$HOCH_2CH_2OH$	-12.90	9.62

Table 1. Molal Freezing-Point Depression Constants for Several Substances

Let's use **Equation 1** to determine the freezing point of a solution made of 40.0 g of ethanol, C_2H_5OH, dissolved in 60.0 g of water. We must first determine the molal concentration of the solution.

$$\text{molar mass of ethanol} = 46.0 \text{ g/mol}$$

$$\text{moles of ethanol} = \frac{40.0 \text{ g ethanol}}{46.0 \text{ g/mol ethanol}} = 0.870 \text{ mol ethanol}$$

$$60.0 \text{ g H}_2\text{O} = 0.0600 \text{ kg H}_2\text{O}$$

$$\text{molality of solution} = \frac{0.870 \text{ mol ethanol}}{0.0600 \text{ kg H}_2\text{O}} = 14.5\ m$$

Now that we know the molal concentration of the solution we can use **Equation 1** to determine the freezing-point depression.

$$\Delta T_\text{f} = k_\text{f} m$$

$$\text{For H}_2\text{O},\ k_\text{f} = 1.86\ \text{°C}/m$$

$$\text{For this solution molal concentration} = 14.5\ m$$

$$\Delta T_\text{f} = (1.86\ \text{°C}/m)(14.5\ m) = 26.97\ \text{°C} \approx 27.0\ \text{°C}$$

Since water freezes at 0.00 °C this solution will freeze 27.0 °C lower.

Freezing point of solution = -27.0 °C

Procedure

Part A – Determine Freezing Point of Water

1. Using a balance, weigh and record the mass of a large test tube.
2. Pour approximately 25.0 mL of distilled or deionized water into the large test tube.
3. Using the same balance, weigh and record the mass of the water and test tube. Determine by difference the mass of the water.
4. Place one or two cups of crushed ice into a 500-mL beaker. Add 15 to 20 g of NaCl to the ice, along with water, and stir to make a slurry. (Maintain an ice/salt temperature of -5^{0}C . You may have to remake this bath during the experiment.)
5. Place a glass or metal stirring rod into the test tube containing the distilled or deionized water.

6. Place an alcohol thermometer into the test tube containing the distilled or deionized water.
7. Assemble the experimental setup as indicated in **Figure 3**, adding the test tube containing the distilled or deionized water last.
8. Stir the distilled or deionized water using the stirring rod. At 10-second intervals record the time and temperature of the water in your lab manual.
9. Continue to stir the water until it freezes and remains frozen for at least one minute. Do not stop until you are convinced that the freezing temperature remains unchanged.
10. Make a plot of temperature (y-axis) versus time (x-axis) using your recorded data. Your plot should look similar to **Figure 4**.
11. Allow the distilled water to warm to room temperature. This can be accomplished by letting the test tube sit on the bench top or placing the test tube in a beaker containing warm water.
12. Repeat steps 7 through 10 to perform a second trial.
13. Allow the distilled water to warm to room temperature.

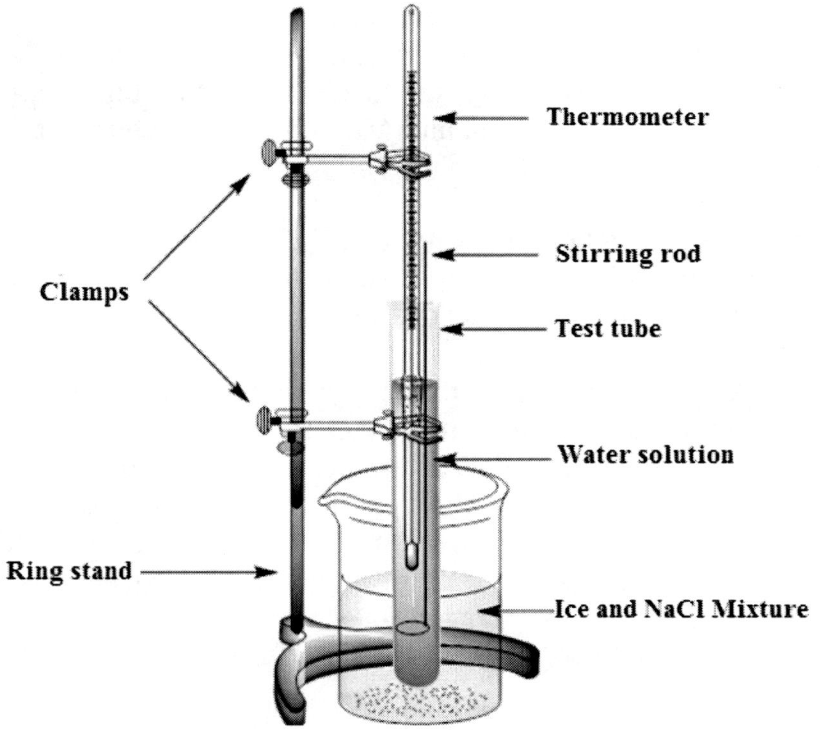

Figure 3. Experimental Setup

Part B – Determine Freezing Point of Water and Ethylene Glycol Mixture

14. Using a balance, weigh and record the mass of a small test tube. Carefully transfer approximately 1.5 mL of ethylene glycol into the test tube. (Be careful not to ingest any of the ethylene glycol.)

15. Weigh and then record the mass of the test tube and ethylene glycol. Determine by difference the mass of the ethylene glycol.
16. Pour the ethylene glycol into the test tube containing the distilled or deionized water.
17. Using the measured mass of the water and ethylene glycol (HO-CH_2-CH_2-OH), determine the molal concentration of the water and ethylene glycol mixture.
18. Return the large test tube containing the water and ethylene glycol mixture to the experimental setup and place it in the beaker containing ice and NaCl.
19. Repeat the stirring and cooling procedure used in Part A, being sure to record the temperature every 10 seconds. Record both the time and temperature in your lab manual.
20. Continue stirring as the mixture cools until the mixture freezes and remains so for at least one minute. Record data until you are convinced that the freezing point temperature remains unchanged.
21. Make a plot of temperature (*y*-axis) versus time (*x*-axis) using your recorded data.
22. Allow the water and ethylene glycol mixture to warm to room temperature. This can be accomplished by letting the test tube sit on the bench top or by placing the test tube in a beaker containing warm water.
23. Repeat steps 19-21 to perform a second trial.
24. Allow the water and ethylene glycol mixture to warm to room temperature.
25. Pour the water and ethylene glycol mixture into an approved waste container.
26. Clean and dry the large test tube.

<u>Part C – Determine Freezing Point of Water and NaCl Mixture</u>

27. Inside the large test tube prepare a distilled water and NaCl mixture that has the same molal concentration as the distilled water and ethylene glycol mixture used in Part B.
28. Return the large test tube containing the water and NaCl mixture to the experimental setup and place it in the beaker containing ice and NaCl.
29. Repeat the stirring and cooling procedure, recording the temperature every 10 seconds. Record both the time and temperature in your lab manual.
30. Continue stirring as the mixture cools until the mixture freezes and remains so for at least one minute. Record data until you are convinced that the freezing point temperature remains unchanged.
31. Make a plot of temperature (*y*-axis) versus time (*x*-axis) using your recorded data.
32. Allow the water and NaCl mixture to warm to room temperature. This can be accomplished by letting the test tube sit on the bench top or placing the test tube in a beaker containing warm water.
33. Repeat steps 29 and 31 to perform a second trial.
34. While the water and NaCl mixture warms, disassemble the experimental apparatus. The ice and NaCl mixture, as well as the water and NaCl mixture, can be poured down the drain.
35. Clean and dry the large test tube, beaker, and thermometer. Return the equipment to your lab drawer or stockroom.

Experiment 16 Pre-Lab (Show ALL Work and ALWAYS include units)

1. Shown below are cooling curves for distilled water and an aqueous solution. Use the data to determine the freezing-point depression of the aqueous solution. Notice that the temperature for both liquids drops below the equilibrium temperature briefly before returning to the final equilibrium temperature. This phenomenon, called supercooling, occurs when the liquid has no particles to form an initial crystal around. As soon as one crystal forms, the liquid quickly freezes, returning to its normal freezing temperature. Stirring or addition of a crystal typically stimulates the solution to freeze.

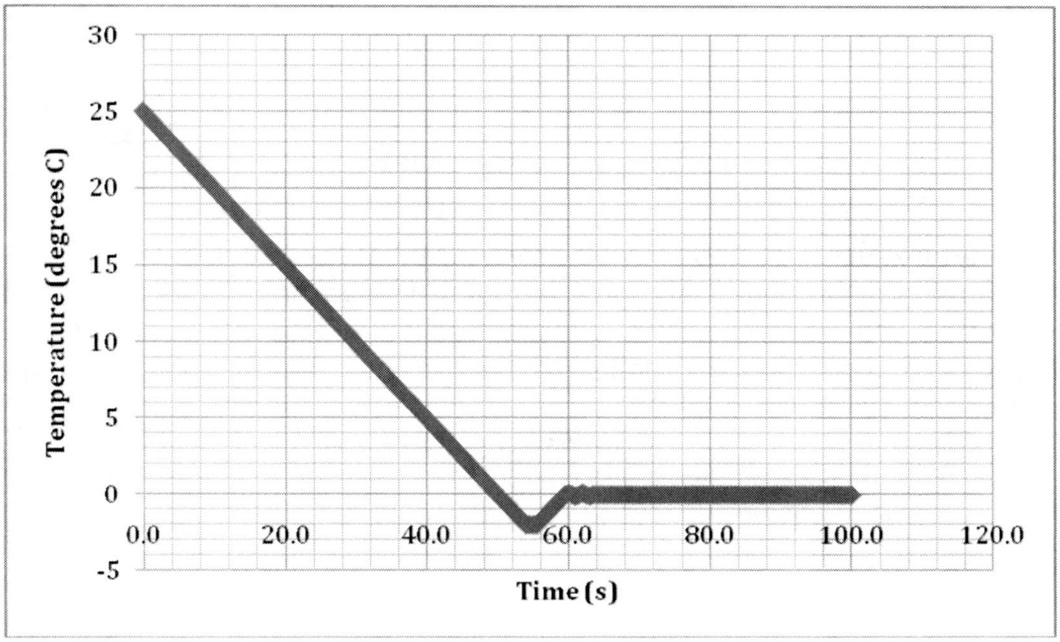

Figure 4. Representative Distilled Water Cooling Curve

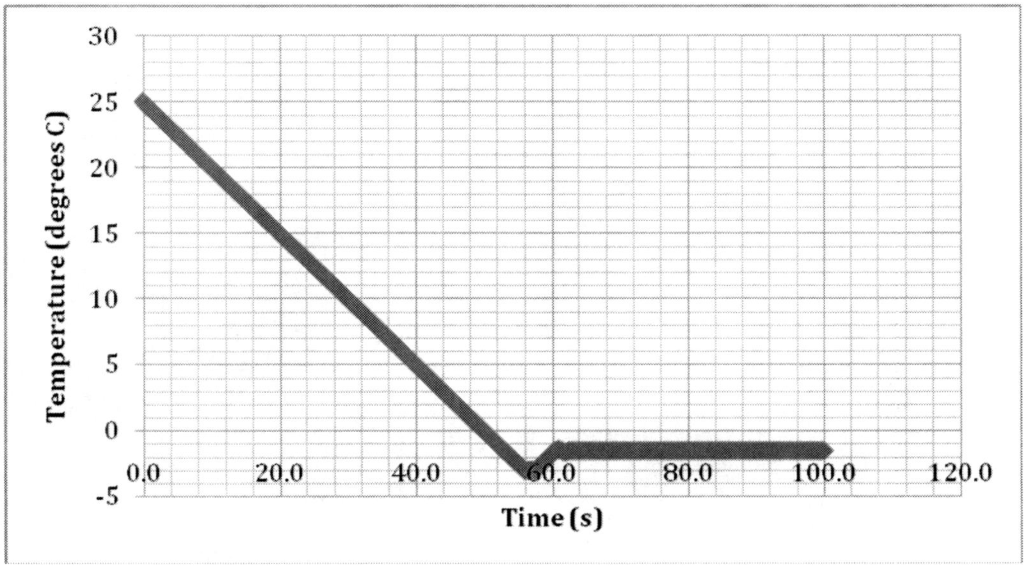

Figure 5. Representative Aqueous Solution Cooling Curve

2. Using the freezing-point depression from the first step and water's k_f value (**Table 1**), determine the molal concentration of the aqueous solution.

3. If the solute used to make the aqueous solution from above has a molar mass of 75.0 g/mole, how many grams of solute are required to dissolve in 24.8 g of water to make a solution having the concentration determined in step 2?

4. To make an aqueous solution having a freezing point of −23.5 °C (−10.3 °F), a 40% by volume ethylene glycol solution is required. If your car has a cooling system with a total volume of 9.5 quarts, how many gallons of ethylene glycol antifreeze do you need to pour into the cooling system to ensure the system does not freeze at temperatures above −10.3 °F?

   4 quarts = 1 gallon

Experiment 16 Lab Report (Show ALL Work and ALWAYS include units)

Part A – Determine Freezing Point of Water

1. Mass of distilled water and test tube ________ g

   Mass of test tube ________ g

   Mass of water ________ g

   1st Cooling Curve

   Record the water temperature (°C) for every 10 seconds of cooling time.

Time	Temp.	Time	Temp.	Time	Temp.	Time	Temp.	Time	Temp.

2nd Cooling Curve

Record the water temperature (°C) for every 10 seconds of cooling time.

Time	Temp.	Time	Temp.	Time	Temp.	Time	Temp.	Time	Temp.

Plot the temperature (y-axis) versus time (x-axis) using your recorded data on a piece of graph paper or using Excel or some similar graphing software for both trials. (Ask your instructor which form is acceptable.)

Part B – Determine Freezing Point of Water and Ethylene Glycol Mixture

2. Mass of distilled water, test tube, and ethylene glycol ________ g

   Mass of distilled water and test tube ________ g

   Mass of ethylene glycol ________ g

Moles of ethylene glycol ________ mol

Molal concentration of water and ethylene glycol mixture ________ molal

1st Cooling Curve of Water and Ethylene Glycol Mixture

Record the mixture temperature (°C) for every 10 seconds of cooling time.

Time	Temp.	Time	Temp.	Time	Temp.	Time	Temp.	Time	Temp.

2nd Cooling Curve of Water and Ethylene Glycol Mixture

Record the water temperature (°C) for every 10 seconds of cooling time.

Time	Temp.	Time	Temp.	Time	Temp.	Time	Temp.	Time	Temp.

Plot the freezing temperature (y-axis) versus time (x-axis) using your recorded data on a piece of graph paper or using Excel or some similar graphing software for both trials (Ask your instructor which form is acceptable.)

Part C – Determine Freezing Point of Water and NaCl Mixture

3. Mass of distilled water and test tube ________ g

   Mass of test tube ________ g

   Mass of water ________ g

   Required molal concentration of water and NaCl mixture ________ molal

   Moles of NaCl required ________ mol

   Mass of NaCl required ________ g

1st Cooling Curve of Water and NaCl Mixture

Record the mixture temperature (°C) for every 10 seconds of cooling time.

Time	Temp.	Time	Temp.	Time	Temp.	Time	Temp.	Time	Temp.

2nd Cooling Curve of Water and NaCl Mixture

Record the water temperature (°C) for every 10 seconds of cooling time.

Time	Temp.	Time	Temp.	Time	Temp.	Time	Temp.	Time	Temp.

Plot the freezing temperature (y-axis) versus time (x-axis) using your recorded data on a piece of graph paper or using Excel or some similar graphing software for both trials. (Ask your instructor which form is acceptable.)

4. From the graphical data determine the freezing-point depression (change in freezing-point temperature) of the water and ethylene glycol mixture.

5. From the graphical data determine the freezing-point depression (change in freezing-point temperature) of the water and NaCl mixture.

6. Explain in your own words why there is a difference between the freezing-point depression of the two solutions that have the same molal concentration.

7. How could you modify the freezing-point depression expression to account for this difference?

Experiment 17
Examining a Chemical Reaction Using Kinetics

Objective

- Use a chemical reaction to study kinetics.

Topics Introduced in this Lab

- Chemical kinetics
- Reaction rates
- Rate laws
- Method of initial rates
- Catalysts

Related Zumdahl Sections

- Chapter 11 Sections 1, 2, 3 and 7

Safety Considerations

- An acid catalyst is used. Acids are corrosive.
- Acetone is an organic solvent and is flammable. No open flames should be used during this experiment.
- Iodine can stain your skin and clothing.
- If any of these chemicals get on you, wash immediately with copious amounts of water followed by soap and water.
- Do not ingest any of these compounds.
- Wear safety glasses and appropriate lab wear at all times while in the lab.

Introduction

Reaction kinetics is a study of the dependence of the reaction rate on the concentration of the species involved in the reaction. Understanding this dependence helps chemists determine how a reaction occurs on the molecular level. For instance, for the generic reaction shown in **Equation 1**, studying the kinetics of the reaction can help answer questions such as "Does A react with B first, and then C?" or "Does B react with C first, and then A?" Initially, these may seem like trivial questions. However, a clear understanding of exactly what is happening in a chemical reaction can lead to increased reaction yields and decreased reaction times. This in turn can lead to more efficient and cost-effective production of crucial pharmaceuticals, materials, etc.

$$\mathbf{A + B + C \rightarrow D}$$

Equation 1. Generic Chemical Reaction

A rate law is a mathematical expression that relates the rate of a reaction to the concentrations of reactants or products. For the generic reaction given above, a possible rate law is shown in **Equation 2** below.

$$\textbf{rate} = k\,[\mathrm{A}]^{a}\,[\mathrm{B}]^{b}\,[\mathrm{C}]^{c}$$

where k = rate constant
[A] = concentration of A
a = order with respect to [A]
and similarly for B, C, b, and c

Equation 2. Rate Law for Generic Reaction

In this rate law, "k" is the rate constant for the reaction and is unique for a given reaction at a specific temperature. The letters "a", "b" and "c" are the orders of the reaction with respect to the reactant concentrations [A], [B], and [C], respectively, and describe how the reaction's rate depends on each of those reactant concentrations. One goal of chemical kinetics is to "fill-in" the values for the rate constant and the orders, thereby completing the rate law and developing a clear picture of how a given chemical reaction occurs.

In this experiment, you will determine the rate constant and the orders for the reaction of iodine with acetone in the presence of an acid catalyst as shown in **Equation 3**.

$$CH_3COCH_3(aq) + I_2(aq) \xrightarrow{H^+(aq)} CH_3COCH_2I(aq) + H^+(aq) + I^-(aq)$$

Equation 3. Reaction of Acetone with Iodine

To accomplish this, you will use a kinetic method known as the method of initial rates. This method relies on measuring the rate at the initial stages of a reaction for multiple reaction trials, while the reactant concentrations are systematically varied one at a time. By altering only one concentration at a time and observing the effect on the rate of the reaction, the reaction's rate dependence on that reactant is determined. For example, a student studying the kinetics of the generic reaction shown in **Equation 1** might run multiple reaction trials with the reactant amounts indicated in **Table 1**.

Rxn	mL of A [1.0 *M*]	mL of B [1.0 *M*]	mL of C [1.0 *M*]	mL of H_2O	Initial Rate (*M*/s)
#1	10.0 mL	10.0 mL	10.0 mL	10.0 mL	0.010
#2	20.0 mL	10.0 mL	10.0 mL	0.0 mL	0.040
#3	10.0 mL	20.0 mL	10.0 mL	0.0 mL	0.020
#4	10.0 mL	10.0 mL	20.0 mL	0 0 mL	0.010

Table 1. Sample Initial Rates Study

Examine reaction trials 1 and 2 from **Table 1**. Note that only two things are changed between the two trials, the volume of A and the volume of water used. The volume of A is doubled in trial 2. This changes the concentration of A in the reaction mixture, which also affects the reaction rate. However, if the water volume (which is our solvent in this case) was not changed as well, then the total reaction mixture volume would have been 50.0 mL in trial 2 instead of 40.0 mL as it was in trial 1. This change in total reaction volume would cause the concentrations of the other reactants (B and C) to change in addition to reactant A. To isolate and determine the reaction rate dependence on a specific reactant, ONLY that reactant's concentration can be changed from one trial to the next.

Now let's take a more detailed look at reaction trial 1 and reaction trial 2 in **Table 1**. As the concentration of A doubles the concentrations of B and C are kept constant and the reaction rate quadruples. This reaction rate change can only be due to the change in the concentration of A since all other concentrations were kept constant between trials. This implies that the rate is proportional to the square of the concentration of A. In other words, the reaction is second order with respect to reactant A (a = 2). This is expressed mathematically as shown in **Equation 4**.

$$\frac{\boldsymbol{Rate\ Law\ for\ Trial\ 1}}{\boldsymbol{Rate\ Law\ for\ Trial\ 2}}$$

$$\frac{\boldsymbol{rate_1}}{\boldsymbol{rate_2}} = \frac{[\mathrm{A}]^a[\mathrm{B}]^b[\mathrm{C}]^c}{[\mathrm{A}]^a[\mathrm{B}]^b[\mathrm{C}]^c}$$

$$\frac{0.010\ M/s}{0.040\ M/s} = \frac{(0.25M)^a(0.25M)^b(0.25M)^c}{(0.50M)^a(0.25M)^b(0.25M)^c}$$

$$\frac{0.010\ M/s}{0.040\ M/s} = \frac{(0.25M)^a}{(0.50M)^a}$$

$$0.25 = (0.50)^a$$

$$a = 2.0$$

Equation 4. Mathematical Determination of Reaction Order with Respect to A

Now look at trials 1 and 3 in **Table 1**. The concentration of B doubles while the concentration of everything else remains the same. The reaction rate doubles from trial 1 to trial 3. Since the concentration of B doubles and the reaction rate doubles, the reaction rate is proportional to the concentration of B. Therefore, the reaction is first order with respect to B (b = 1).

Similarly examine reaction trials 1 and 4 in **Table 1**. The concentration of reactant C doubles while the concentration of everything else remains constant. The rate of the reaction stays the same from trial 1 to trial 4, however. This indicates that there is no dependence of the reaction rate on the concentration of C. The reaction is zeroth order with respect to reactant C ($c = 0$).

The rate law for this general reaction is now partially filled in:

$$\textbf{rate} = k\,[\mathrm{A}]^2\,[\mathrm{B}]^1\,[\mathrm{C}]^0$$

Equation 5. Rate Law for Generic Reaction – Partially Completed

To determine the value of k and complete the rate law, we can use a set of data from one of the reaction trials. In your experiment, you will calculate a value for k from each reaction trial and then average those k values together. For this example, we'll just determine the value of k based on the data from only one trial. Let's choose the data from trial 3 to determine the value of k, using **Equation 6** below.

$$\textbf{rate} = k\,[\mathrm{A}]^2\,[\mathrm{B}]^1\,[\mathrm{C}]^0$$

$$0.020\ M/\mathrm{s} = k(0.25\ M)^2(0.50\ M)^1(0.25\ M)^0$$

$$0.020 = k(0.031)$$

$$k = 0.65\ \frac{1}{M^2\mathrm{s}}$$

Equation 6. Determining the Value of k

Now we have enough information to complete the rate law for this reaction. The final version is shown in **Equation 7**.

$$\textbf{rate} = 0.65\ \frac{1}{M^2\mathrm{s}}\ [\mathrm{A}]^2\,[\mathrm{B}]^1\,[\mathrm{C}]^0$$

Equation 7. Completed Rate Law for Generic Reaction

In this experiment, you'll use the initial rates method as illustrated for our generic reaction above, except you'll be examining the reaction of iodine with acetone as shown in **Equation 3**.

A significant question still remains unanswered: How do we measure the rate of a reaction? Various analytical techniques can be used to monitor the disappearance of reactants over time for chemical reactions, thereby allowing for the determination of the reaction rate. These techniques range from very simple to extremely complex methods requiring advanced analytical equipment. The reaction in this experiment has a colored reactant and only colorless products. As the reaction proceeds, the yellow color of iodine will slowly disappear as the iodine reacts with the acetone. Due to this characteristic, the rate of the reaction is easily determined visually by measuring the time it takes for the iodine's color to disappear. The rate is then the concentration change of the iodine over time. This is illustrated in **Equation 8**. Note that the rate equals the *negative* of the change in concentration of iodine over time since iodine is being consumed in the reaction, and the reaction progresses forward as the concentration of iodine decreases.

$$rate = \frac{-\Delta[I_2]}{\Delta t} = \frac{-([I_2]_{final} - [I_2]_{initial})}{t_{final} - t_{initial}} = \frac{-(0 - [I_2]_{initial})}{t_{final} - 0} = \frac{[I_2]_{initial}}{t_{final}}$$

Equation 8. Determining the Rate for the Reaction of Iodine with Acetone

For this experiment, you will systematically vary the concentrations of the reactants (iodine and acetone) as well as the catalyst (H^+) for the reaction shown in **Equation 3**. For each reaction trial (where the concentration of only one component is changed), you will calculate the rate of the reaction. Comparing the concentrations and the rate determined for select reaction trials, you will calculate the order of the reaction with respect to each component (iodine, acetone, and H^+). Then, you will calculate the rate constant (k) for the reaction using the concentration, rate, and order information from each of the reaction trials. The orders and rate constant will then be used to write the complete rate law for the reaction of acetone (CH_3COCH_3) with iodine in the presence of an acid catalyst. The incomplete form of this rate law is shown in **Equation 9**.

$$\textbf{rate} = k\,[I_2]^a\,[CH_3COCH_3]^b\,[H^+]^c$$

Equation 9. Incomplete Rate Law for Reaction of Iodine with Acetone

Procedure (*Be sure to record all relevant data in your lab report.)

Rxn	mL of 0.0010 M I_2	mL of 4.0 M Acetone	mL of 0.10 M HCl	mL of distilled or deionized H_2O
#1	1.0 mL	1.0 mL	1.0 mL	2.0 mL
#2				
#3				
#4				

Table 2. Reactant and Solvent Amounts for Experimental Trials.

1. Obtain 5 large, clean test tubes and a test tube rack. Label the test tubes as 0 through 4.
2. In test tube #0, put 5.0 mL of distilled or deionized water. This test tube will serve as a reference to help you determine when all color has disappeared from your reaction mixtures.

3. **Table 2** shows the amount of each reactant, catalyst, and solvent to be used in reaction #1. Using the values for reaction #1 as a starting point, complete **Table 2** by determining the amounts of reactants, catalyst, and solvent you should use in reactions #2–4, to systematically double the concentration of only one of the reactants or the catalyst at a time for each trial. Remember that the total volume of the reaction mixture should remain constant for all reactions. Have your instructor double check your planned volumes for reactions #2–4 before proceeding.
4. In test tube #1, put the amounts of reagents listed for Rxn #1, **Table 2** EXCEPT for the iodine. (Test tubes 1–4 will all be prepared without iodine initially. Once all test tubes have been prepared, iodine will be added quickly in the appropriate amounts to all test tubes to initiate the reaction nearly simultaneously for all trials.)
5. Prepare test tubes 2–4 using your planned volumes from Rxn #2-4, **Table 2** respectively, keeping in mind that you should not add the iodine yet.
   *Make sure that a clock is available so that you can time your reactions in seconds.
6. Measure out the appropriate volume of iodine that you will need for test tubes 1–4. With these volumes measured out, quickly add the appropriate volume of iodine to each test tube and begin timing the reactions. Quickly swirl the contents of each test tube briefly to ensure proper mixing before replacing the test tubes in the test tube rack.
7. Record the time that it takes for each reaction mixture to become colorless, using test tube 0 as a reference. (A piece of white paper held behind the test tubes can make the color change more apparent.)
8. After all reactions have finished, empty the contents of your test tubes, except for test tube #0, into the appropriate waste container. Clean and dry your test tubes.
9. Repeat steps 4–8 to perform 2 more trials of each reaction in **Table 2**.

Experiment 17 Pre-Lab (Show ALL Work and ALWAYS include units)

Assume you collected the following data in lab while studying the generic reaction

$$A + B + C \rightarrow D + E$$

Sample Data

Rxn	mL of A [0.10 *M*]	mL of B [0.20 *M*]	mL of C [0.20 *M*]	mL of H_2O	Initial Rate (*M*/s)
#1	1.0 mL	2.0 mL	2.0 mL	5.0 mL	0.0011
#2	1.0 mL	2.0 mL	4.0 mL	3.0 mL	0.0040
#3	1.0 mL	4.0 mL	4.0 mL	1.0 mL	0.0039
#4	2.0 mL	2.0 mL	2.0 mL	4.0 mL	0.0021

1. Determine the order of the reaction with respect to reactant A. You must either show a calculation or give a detailed written explanation of your answer.

2. Determine the order of the reaction with respect to reactant B. You must either show a calculation or give a detailed written explanation of your answer.

3. Determine the order of the reaction with respect to reactant C. You must either show a calculation or give a detailed written explanation of your answer.

4. In this pre-lab example, you are only looking at the initial rate of the reaction, that is, the rate at the very beginning of the reaction before any significant change has occurred in the concentrations of the reactants. In other words, the reactant concentrations can be treated as constant during the time that the reaction rate is measured. This is an important requirement when using the method of initial rates to study the kinetics of a reaction. If you were to measure the rate throughout the course of a reaction, the rate would constantly be changing as the reactant concentrations are depleted. By examining only the instantaneous initial rate, the cause of any changes in rates between trials can be correlated directly to the changes in initial concentrations used between those trials, rather than changes that occur during the course of the reactions. If you have carefully read the procedure for this experiment, you know that you will be determining the rate by timing the entire reaction from start to finish, NOT just at the beginning. This would seem to be in direct contradiction to the method of initial rates, which we are using. However, two special conditions allow for this "modification" of the method. First, the reaction is known to be zero order with respect to iodine (which you will confirm in this experiment). Second, the concentrations of acetone and H+ catalyst used are very large compared to that of the iodine that is used. Using these two conditions, explain how we can accurately study the reaction kinetics using the method of initial rates while measuring the reaction rate over the course of the entire reaction instead of only the initial rate.

Experiment 17 Lab Report (Show ALL Work and ALWAYS include units)

1. Complete this table.

Rxn	mL of 0.0010 *M* I_2	mL of 4 . 0 *M* Acetone	mL of 0.10 *M* HCl	mL of H_2O
#1	1.0 mL	1.0 mL	1.0 mL	2.0 mL
#2				
#3				
#4				

2. Complete this table for the initial concentrations in the **reaction** mixtures.

Rxn	$[I_2]_{initial}$	$[Acetone]_{initial}$	$[H^+]_{initial}$
#1			
#2			
#3			
#4			

3. Show a sample calculation for the determination of $[I_2]_{initial}$ in the reaction mixture for one of your reaction trials.

4. Complete this table (use the average reaction time to calculate the reaction rate).

Rxn	Trial	Reaction Time (s)	Average Reaction Time (s)	Reaction Rate (*M*/s)
#1	1			
	2			
	3			
#2	1			
	2			
	3			
#3	1			
	2			
	3			
#4	1			
	2			
	3			

5. Show a sample calculation for the determination of the reaction rate for one of your reaction trials.

6. Which two reactions should you compare to determine the order of the reaction with respect to iodine?

7. Determine the order of the reaction with respect to iodine. You must either show a calculation or give a detailed written explanation of your answer.

8. Which two reactions should you compare to determine the order of the reaction with respect to acetone?

9. Determine the order of the reaction with respect to acetone. You must either show a calculation or give a detailed written explanation of your answer.

10. Which two reactions should you compare to determine the order of the reaction with respect to H^+?

11. Determine the order of the reaction with respect to H^+. You must either show a calculation or give a detailed written explanation of your answer.

12. Complete this table. (Use the rates calculated in Question #4 for the first column of this table. The initial concentrations are the same as those in Question #2 above.)

Rxn	Reaction Rate (M/s)	$[I_2]_{initial}$	$[Acetone]_{initial}$	$[H^+]_{initial}$	Rate Constant (k)	Average Rate Constant (k)
#1						
#2						
#3						
#4						

13. Show a sample calculation for determining the rate constant (k) using the data from one of your reaction trials.

14. Write the complete rate law for the reaction of iodine with acetone in the presence of an acid catalyst using the orders you determined for the reactants and catalyst and the average rate constant (k) that you determined in Question #12.

Experiment 18
NO_2 – N_2O_4 Equilibrium –
Le Châtelier's Principle in Action

Objectives

- Generate NO_2 gas from copper and nitric acid.
- Examine the equilibrium established between NO_2 and N_2O_4.
- Determine the effect of temperature change on the NO_2 – N_2O_4 equilibrium.

Topics Introduced in this Lab

- Le Chatelier's principle
- Equilibrium

Related Zumdahl Sections

- Chapter 8 Sections 3, 4, and 10
- Chapter 12 Sections 1, 2, 5, and 7

Safety Considerations

- Concentrated nitric acid is very corrosive and damaging to skin and eyes. Be careful with this substance when using or pouring it. Handle with care.
- NO_2 gas is toxic. As the gas is being generated you must have adequate ventilation. It is best to produce the gas inside a fume hood or other similar ventilation device.
- If any of these compounds come into contact with your eyes or skin, immediately flush the affected area with water.
- Do not ingest any of these compounds.
- Wear safety glasses and appropriate lab wear at all times while in the lab.

Introduction

Internal combustion engines, such as those in a car or lawn mower engine, burn a fuel, typically gasoline, in compressed air. Because air is essentially 80% N_2 and 20% O_2, one of the by-products of combustion is nitrogen dioxide, NO_2, produced as shown in **Equation 1.**

$$N_2(g) + 2O_2(g) \rightarrow 2NO_2(g)$$

Equation 1. Reaction of Nitrogen with Oxygen to Form Nitrogen Dioxide

This reaction only occurs in significant yields at high temperatures, which is fortunate, because nitrogen dioxide is toxic. If the reaction occurred at normal temperatures we would die from its production in air. Nitrogen dioxide is a significant component of air pollution near large metropolitan areas where automobile traffic is heavy. Nitrogen dioxide is a brown-colored gas that typically rises in the atmosphere to a few hundred feet above ground level. When taking off from major metropolitan airports, like Los Angeles or Atlanta, in the summer, a distinct nitrogen dioxide layer is visible outside the airplane windows.

Nitrogen dioxide has 17 electrons. Consequently, it has one unpaired electron. Molecules containing an unpaired electron are called free radicals. Free radicals are quite reactive, which partially explains the toxicity of nitrogen dioxide. It can also dimerize, react with itself, to form dinitrogen tetroxide, **Equation 2**. Dinitrogen tetroxide is red colored. As the equilibrium described in **Equation 2** is established, the mixture acquires a visible reddish-brown color.

$$2NO_2(g) \rightleftharpoons N_2O_4(g)$$

Equation 2. Dimerization of Nitrogen Dioxide to Form Dinitrogen Tetroxide

At equilibrium the forward and reverse reaction rates are equal. Forward and reverse reaction rates are influenced by temperature and concentration changes. For example, changing reactant concentrations increases the forward reaction rate relative to the reverse reaction rate, resulting in increasing product concentration. Likewise, if the product concentration is increased the reverse reaction rate also increases, resulting in increased reactant concentration. The French chemist Henry Louis Le Châtelier explained this phenomenon by stating that when stressed, a system at equilibrium shifts in the direction that reduces that stress. This phenomenon is called Le Châtelier's principle.

Temperature changes also stress a system at equilibrium. For exothermic reactions at equilibrium, a temperature increase *decreases* product concentration and *increases* reactant concentration. Cooling an exothermic reaction by decreasing the equilibrium temperature has the opposite effect: increasing product concentration and decreasing reactant concentration. Endothermic equilibrium reaction systems behave oppositely to exothermic systems as indicated in **Table 1**.

Equilibrium Reaction System	Temperature Change	Reactant Concentration	Product Concentration
Exothermic	Increase	Increases	Decreases
Exothermic	Decrease	Decreases	Increases
Endothermic	Increase	Decreases	Increases
Endothermic	Decrease	Increases	Decreases

Table 1. Equilibrium Reactions Response to Temperature Stress

This experiment employs absorption spectroscopy to monitor the changes in equilibrium concentrations. If you know the absorbance value for $NO_2(g)$ at room temperature then heat or cool the gas, any detectable change in absorbance reflects an $NO_2(g)$ concentration change. A plot of absorbance values versus time reflects how the NO_2 – N_2O_4 equilibrium responds to changes in temperature.

In this experiment you will generate $NO_2(g)$ via the reaction of nitric acid with copper, **Equation 3**. You will fill a cuvette or test tube with $NO_2(g)$ and cap it with either a top or piece of tape. Next, you will take a visible absorption spectrum of $NO_2(g)$ to determine the wavelength appropriate to monitor concentration changes. Then you will stress the system by heating and cooling the cuvette. By monitoring the absorption spectrum as a function of time after heating and cooling, you can see the system respond to stress as well as determine if the reaction is endothermic or exothermic.

$$\mathbf{Cu}(s) + \mathbf{4HNO_3}(aq) \rightarrow \mathbf{Cu(NO_3)_2}(aq) + \mathbf{2NO_2}(g) + \mathbf{2H_2O}(l)$$

Equation 3. Reaction of Cu and Nitric Acid

Procedure

Part A – Generate $NO_2(g)$

1. Obtain 5.0 to 10.0 g of copper metal pieces, 150.0 mL of 6.0 *M* $HNO_3(aq)$, and a test tube or cuvette which fits into the absorption spectrophotometer available in the lab.
2. Tightly attach plastic or rubber tubing to the side-arm nipple of a 500-mL filter flask, **Figure 1**.
3. Place the copper metal into the filter flask.
4. Move the experimental apparatus to a fume hood, fume extractor, or other safe area for handling noxious fumes. **Do not pour the nitric acid into the filter flask until you are in a safe area for handling noxious fumes.**
5. Pour the 150 mL of nitric acid into the filter flask. Quickly place a tight fitting rubber stopper into the filter flask top.
6. Allow the nitric acid to react with the copper for 3 or 4 minutes. Air space above the nitric acid will first fill with brownish colored $NO_2(g)$ then begin flowing out the plastic or rubber tubing.
7. Wearing plastic gloves, place the tubing into the test tube or cuvette opening, completely filling it with $NO_2(g)$ while displacing all the air. Once full, remove the test tube or cuvette and cap it with a stopper, cuvette top, or tape as indicated by your instructor. Seal the test tube or cuvette top with some tape but do not extend tape far enough down the side to interfere with the spectrophotometer beam.
8. Once you are certain that your test tube or cuvette is filled and sealed, remove the rubber stopper from the filter flask, allowing any remaining $NO_2(g)$ to escape and be removed by the fume hood or fume extractor.
9. Dispose of the contents of the filter flask into an approved waste container as directed by your instructor.

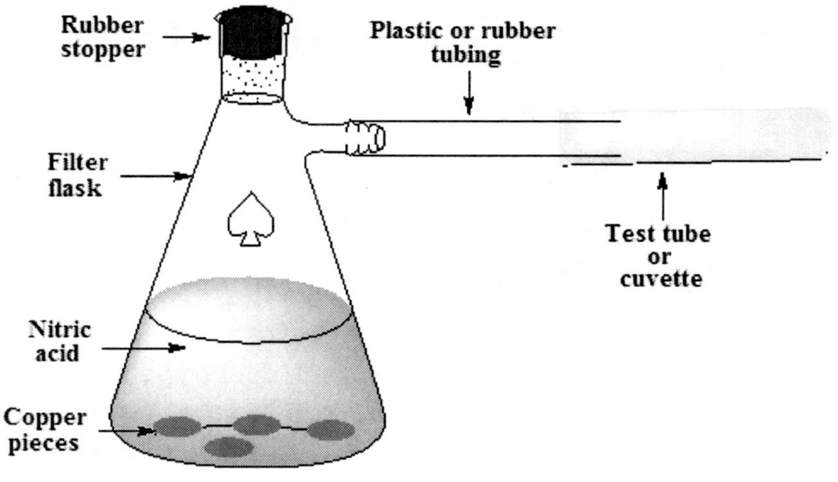

Figure 1. Assembled Experimental Apparatus

Part B – Determine Absorbance Spectrum of $NO_2(g)$

10. Your instructor will give you instructions on how to operate the spectrophotometer in your lab. Be sure to calibrate the spectrophotometer to 0 and 100% transmittance. Typically there is a blackened test tube or cuvette to use for 0% transmittance. A clear cuvette or test tube filled with air is appropriate for 100% transmittance.
11. After the spectrophotometer has been adjusted, place the sealed test tube or cuvette into the spectrophotometer sample chamber. Scan the full wavelength range possible for the spectrophotometer, recording the absorbance at each wavelength. If your lab has manual Spectronic 20's it may be necessary to record the absorbance every 10 nm for the sake of time.
12. As you are measuring the absorbance spectrum of $NO_2(g)$, place 50 mL of water in a 100-mL beaker, then heat the water to 60 °C using a hot plate. In a separate 100-mL beaker place 50 mL of ice.
13. Make a graph of absorbance versus wavelength using your recorded data. Based upon this graph, determine the wavelength where $NO_2(g)$ has the maximum absorbance. (This wavelength, referred to as λ_{max} (lambda max), provides the greatest sensitivity to $NO_2(g)$ concentration changes.)
14. Set the spectrophotometer to the λ_{max} for $NO_2(g)$ and maintain it at this wavelength for the rest of the experiment.

Part C – Measure the Absorbance Spectrum of $NO_2(g)$ That Has Been Warmed

15. Remove the cuvette or test tube from the spectrophotometer. Place it into the beaker containing 60 °C water for approximately 3 minutes, allowing the gas to attain thermal equilibrium with the water.
16. Using tongs or tweezers, remove the cuvette or test tube from the hot water, wipe off any exterior moisture, and place it back into the spectrophotometer. Remember, the spectrophotometer wavelength must still be set at the λ_{max} for $NO_2(g)$.

17. Record the absorbance value at the λ_{max} for $NO_2(g)$ every 10 seconds as the gas mixture cools from 60 °C to room temperature.
18. Repeat this measurement, if necessary, to achieve consistent data.

Part D – Measure the Absorbance Spectrum of $NO_2(g)$ That Has Been Cooled

19. Remove the cuvette or test tube from the spectrophotometer. Place it into the beaker containing ice water for approximately 3 minutes, allowing the gas to attain thermal equilibrium with the water.
20. Remove the cuvette or test tube from the ice water, wipe off any exterior moisture, and place it back into the spectrophotometer. Remember, the spectrophotometer wavelength must still be set at the λ_{max} for $NO_2(g)$.
21. Record the absorbance value at the λ_{max} for $NO_2(g)$ every 10 seconds, as the gas mixture warms from 0 °C to room temperature.
22. Repeat this measurement, if necessary, to achieve consistent data.
23. Place your cuvette or test tube in the fume hood, fume extractor, or other safe area and remove the top. Allow the $NO_2(g)$ to withdraw from the container.
24. On one sheet of graph paper plot the absorbance versus time data you collected in steps 17 and 21.

Experiment 18 Pre-Lab (Show ALL Work and ALWAYS include units)

1. Determine the $NO_2(g)$ volume generated when 4.25 g of copper metal reacts with 150.0 mL of 6.0 *M* $HNO_3(aq)$ at 22.5 °C and 0.875 atm.

2. If **Equation 2** is an endothermic reaction as written, sketch what the graph of absorbance versus time will look like as the mixture cools from 60 °C to room temperature.

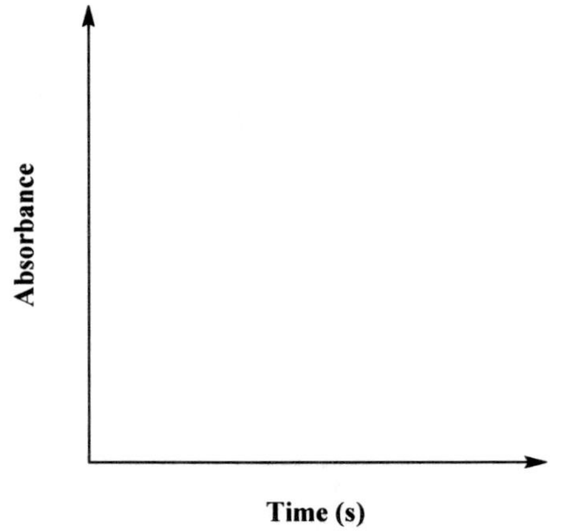

3. If **Equation 2** is an exothermic reaction as written, sketch what the graph of absorbance versus time will look like as the mixture cools from 60 °C to room temperature.

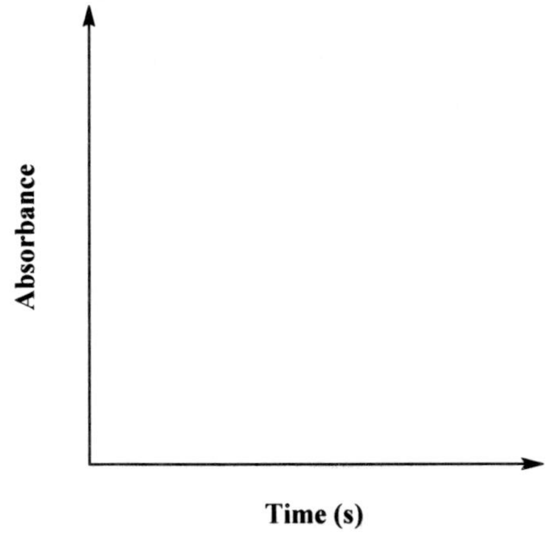

4. Explain your answers to questions 2 and 3, then sketch what the graph of absorbance versus time will look like as the mixture warms from 0 °C to room temperature if **Equation 2** is an endothermic reaction as written.

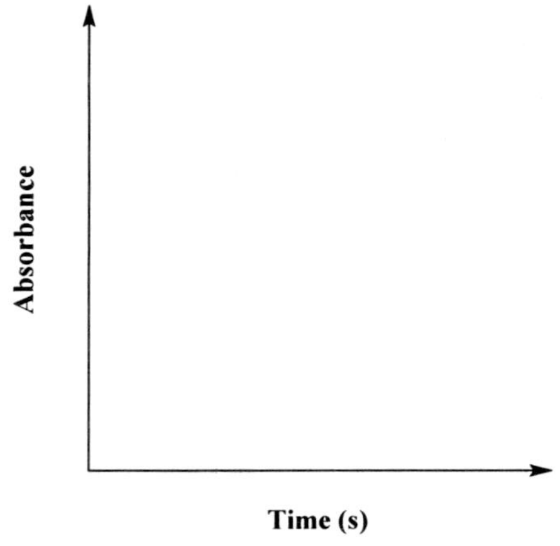

Experiment 18 Lab Report (Show ALL Work and ALWAYS include units)

Part A – Generate $NO_2(g)$

1. What color is the $HNO_3(aq)$ prior to adding it to the copper metal?

2. What color is the aqueous solution after adding it to the copper metal?

3. Explain what caused the color change. Refer to **Equation 3** for help, if necessary.

4. What color is the $NO_2(g)$?

Part B – Determine Absorbance Spectrum of $NO_2(g)$

5. Tape your graph of absorbance versus wavelength here.

6. Based upon your graph from Part B question 1, what is λ_{max} for $NO_2(g)$? Explain how you came to that conclusion.

Part C – Measure the Absorbance Spectrum of $NO_2(g)$ That Has Been Warmed

7. Tape your graph of absorbance versus time for the warmed $NO_2(g)$ here.

8. Explain in your own words what this graph indicates about how the NO_2 – N_2O_4 equilibrium behaves under the stress of being heated.

Part D – Measure the Absorbance Spectrum of $NO_2(g)$ That Has Been Cooled

9. Tape your graph of absorbance versus time for the cooled $NO_2(g)$ here.

10. Explain in your own words what this graph indicates about how the NO_2 – N_2O_4 equilibrium behaves under the stress of being cooled.

11. Based upon your observations, is the NO_2 – N_2O_4 equilibrium as written in **Equation 2** an exothermic or endothermic reaction? Explain your conclusion.

Experiment 19
pH of Various Substances –
Determining Acidic and Basic Properties of Some Substances

Objectives

- Measure the pH of a variety of solutions.
- Understand how pH relates to the acidic or basic nature of chemical substances.

Topics Introduced in this Lab

- pH and pOH
- Acid strength
- Base strength
- Solution concentration

Related Zumdahl Sections

- Chapter 6 Section 3
- Chapter 13 Sections 1, 2, and 3

Safety Considerations

- Concentrated solutions of HCl, NaOH, NH_3, and CH_3COOH can be quite corrosive. Also, they will react with skin and eyes. Handle with care.
- Liquid Plumber® is very corrosive and must be handled with great care. Do not get any of this solution on your skin or in your eyes. Do not mix Liquid Plumber with any other reagents in this experiment.
- If any of these compounds come into contact with your eyes or skin, immediately flush the affected area with water.
- Do not ingest any of these compounds.
- Wear safety glasses and appropriate lab wear at all times while in the lab.

Introduction

One of the most pervasive concepts in chemistry is the idea of acids and bases. While there are several ways to define acids and bases, for this experiment we shall utilize one of the most basic concepts for our definition. Acids are species called electrophiles (electron loving) that seek negative regions of charge. Bases are chemical species called nucleophiles (positive loving) that seek positive regions of charge.

Electrophiles (acids) are either positively charged species or electron deficient. For example, H^+, the hydronium ion, is a very common electrophile in water. It makes sense that a positively charged species would seek electrons to neutralize its charge. However the neutral species, BF_3, is also an electrophile. BF_3 is an electrophile because only 6 valence electrons surround the B atom in BF_3. It is electron deficient. In some reactions BF_3 reacts with electron rich species, F^- for example, to complete its octet of electrons on the B atom, see **Equation 1**. In this experiment we will focus on the positively charged H^+ as our electrophile.

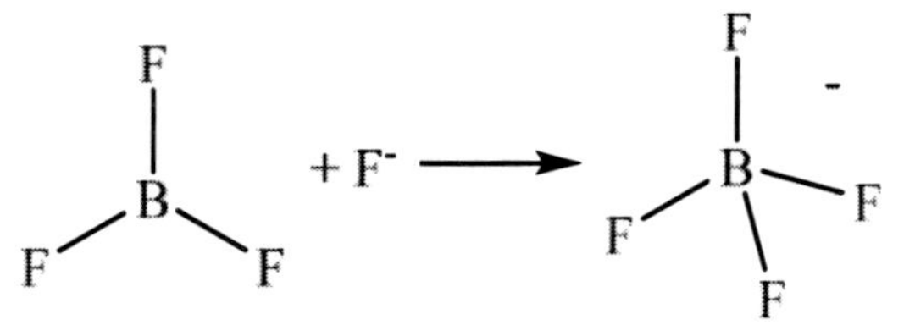

Equation 1. Reaction of BF_3 with F^-

Conversely, nucleophiles are either negatively charged or electron rich species. An excellent aqueous nucleophile is the OH^-, or hydroxide, ion. Being negatively charged, the hydroxide ion seeks positively charged regions on molecules to neutralize its charge. Double and triple bonds, regions of high electron density, are also nucleophiles. For example, ethene readily reacts with H^+ to form a positively charged species, see **Equation 2**. This experiment focuses on the negatively charged hydroxide ion as the nucleophile.

$$H_2C{=}CH_2 + H^+ \rightarrow H_3C{-}CH_2^+$$

Equation 2. Reaction of Hydronium Ion with Ethene

In pure water an equilibrium is established between water and the H^+ and OH^- ions, see **Equation 3**. At 25 °C the molar concentrations of H^+ and OH^- in pure water both equal 1.00×10^{-7} M.

$$H_2O \rightleftharpoons H^+ + OH^-$$

Equation 3. Water, Hydronium, Hydroxide Ion Equilibrium

Using **Equation 4**, we define an equilibrium constant for the water, hydronium and hydroxide ion equilibrium.

$$K_w = \left[H^+\right]\left[OH^-\right] = \left(1.00 \times 10^{-7}\right)\left(1.00 \times 10^{-7}\right) = 1.00 \times 10^{-14} \text{ at } 25^\circ C$$

where $[H^+]$ represents the molar concentration of H^+ ions
$[OH^-]$ represents the molar concentration of OH^- ions

Equation 4. Water Self Ionization Equilibrium Expression

Equation 4 provides a valuable piece of information about this equilibrium. As $[H^+]$ increases, the $[OH^-]$ must decrease to maintain the K_w value. For example, if

$[H^+]$ rises to 1.00×10^{-4}, then the corresponding $[OH^-]$ is easily calculated, see **Equation 5.**

$$K_w = 1.00 \times 10^{-14} = (1.00 \times 10^{-4})[OH^-]$$
$$[OH^-] = \frac{1.00 \times 10^{-14}}{1.00 \times 10^{-4}} = 1.00 \times 10^{-10}$$

Equation 5. Calculation of OH^- Concentration

We define two simple but useful functions in **Equation 6.**

$$pH = -\log[H^+] \text{ and } pOH = -\log[OH^-]$$

Equation 6. Definition of pH and pOH

Using those functions we can simplify **Equation 4** to **Equation 7.**

$$-\log([H^+][OH^-]) = -\log(1.00 \times 10^{-14})$$
$$-\log[H^+] + -\log[OH^-] = -\log(1.00 \times 10^{-14})$$
$$pH + pOH = 14.00$$

Equation 7. pH and pOH Relationship

Equation 7 is another version of **Equation 4.** If we know the pH we also know the pOH. pH and pOH are very simple and useful indicators of acid or base strength. As shown in **Figure 1,** highly acidic solutions have a pH of 1 or 2 and a pOH of 13 or 14. Neutral solutions have a pH = pOH = 7. Very basic solutions have a pH of 13 or 14 and a pOH of 0 or 1. Let's use this scale to determine some information about the acidity of human blood which has a pH = 7.40. Using this information we can determine the pOH as well as the $[H^+]$ and $[OH^-]$.

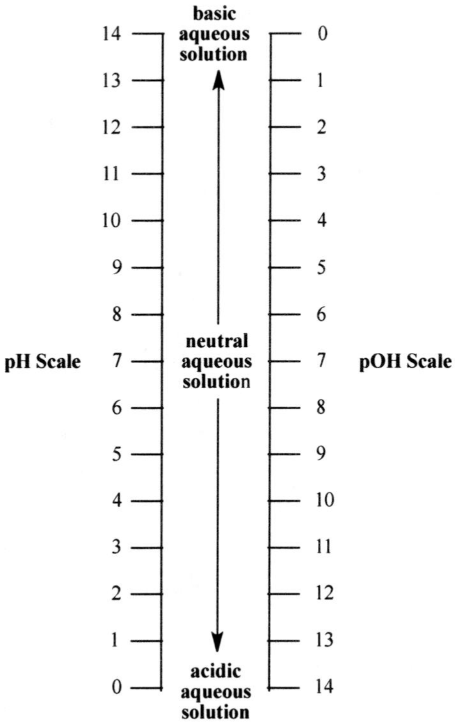

Figure 1. pH and pOH Scales

Because pH + pOH = 14.00, if pH = 7.40 then we can determine that pOH = 14.00 – 7.40 = 6.60. Because pH = –log $[H^+]$, if we take the antilog of the pH, then $[H^+] = 10^{-pH}$. Likewise $[OH^-] = 10^{-pOH}$. Thus, in human blood the $[H^+] = 10^{-7.40} = 3.98 \times 10^{-8}$ *M* and the $[OH^-] = 10^{-6.60} = 2.51 \times 10^{-7}$ *M.* This indicates that human blood is only slightly basic. Indeed it is nearly a neutral solution. However, that slight acidity is important to our bodily functions. It results from the fact that we consume O_2 and exhale CO_2. When dissolved in blood CO_2 forms a slightly basic solution. Among other things, the slight basicity of blood helps to ensure that nutrients and waste flow into and out of cells in the correct direction.

Using a pH probe and meter we will determine the acidity or basicity of lemon juice, Liquid Plumber®, Windex®, Coca Cola®, vinegar, tap water, distilled or deionized water, and a few acid and base solutions.

Procedure

Part A – Determine pH of Several Acid Solutions

1. pH probes consist of an electrode placed into a solution. The whole device is surrounded by glass. At the tip end of the pH probe the glass is bulb shaped and extremely thin. So thin that H^+ ions can pass through the glass. Thus the probes are easily broken if dropped, tapped with a stirring rod, or hit the side of a beaker, so treat them gingerly.

2. When you get your pH probe it will be immersed in a pH = 7 buffer solution or some other storage solution. That is done to keep the electrode solution from drying out. In lab you will also be given two buffer solutions, one acidic and the other basic. After you plug the probe into the pH meter you must wash the probe using a rinse bottle containing distilled or deionized water and dry with a suitable wipe. Place the probe into one of the buffer solutions then adjust the meter readout to the pH of the buffer solution.
3. Remove the pH probe from the buffer solution, then wash it carefully using a rinse bottle containing distilled or deionized water and dry with a suitable wipe. Place the pH probe in the second buffer solution and adjust the meter readout to the pH of the second buffer solution. This gives the pH meter the opportunity to calibrate its electronics so that it will correctly measure pH over the entire range from 0 to 14.
4. Obtain 1.0 *M* HCl(*aq*), 0.50 *M* HCl(*aq*), 0.25 *M* HCl(*aq*), and 0.10 *M* HCl(*aq*) solutions.
5. Pour into a small beaker sufficient 1.0 *M* HCl(*aq*) to cover the bottom one-third of the pH meter.
6. Rinse the pH probe with distilled or deionized water and dry with a suitable wipe. Place the pH meter into the solution. Carefully stir the solution with a glass stirring rod while watching the pH meter. Once the meter has stabilized, record the pH.
7. Repeat the procedure for the 0.50 *M*, 0.25 *M*, and 0.10 *M* HCl(*aq*) solutions.
8. Obtain 1.0 *M* CH_3COOH(*aq*), 0.50 *M* CH_3COOH(*aq*), 0.25 *M* CH_3COOH(*aq*), and 0.10 *M* CH_3COOH(*aq*) solutions.
9. Measure the pH of each CH_3COOH solution using the pH probe and meter. Remember to rinse the pH probe with distilled or deionized water and dry the probe with a suitable wipe in between each measurement. Record the pH measurement of each solution in your lab manual.

Part B – Determine pH of Several Base Solutions

10. Obtain from the stockroom 1.0 *M* NaOH(*aq*), 0.50 *M* NaOH(*aq*), 0.25 *M* NaOH(*aq*), and 0.10 *M* NaOH(*aq*) solutions.
11. Using the same procedure from above, determine the pH of each NaOH solution. Remember to rinse the pH probe with distilled or deionized water in between each measurement. Record the pH measurement of each solution in your lab manual.
12. Obtain from the stockroom 1.0 *M* NH_3(*aq*), 0.50 *M* NH_3(*aq*), 0.25 *M* NH_3(*aq*), and 0.10 *M* NH_3(*aq*) solutions.
13. Repeat the procedure to determine the pH of each NH_3 solution. Remember to rinse the pH probe with distilled or deionized water in between each measurement. Record the pH measurement of each solution in your lab manual.

Part C – Determine pH of Several Common Household Solutions

14. Ask your instructor where to obtain samples of lemon juice, Liquid Plumber®, Windex®, Coca Cola®, vinegar, tap water, and distilled or deionized water.
15. Repeat the procedure you've been using to determine the pH of each common household solution. Remember to rinse the pH probe with distilled or deionized water in between each measurement. Record the pH measurement of each solution in your lab manual.

Experiment 19 Pre-Lab (Show ALL Work and ALWAYS include units)

1. The soft drink Mt. Dew® has a pH of 3.22. Determine the pOH, $[H^+]$, and $[OH^-]$ of Mt. Dew®. Is Mt. Dew® an acidic or basic substance? Does Mt. Dew® contain an electrophile or a nucleophile?

2. When dissolved in water, a typical hand soap has a pOH of 3.75. Determine the pH, $[H^+]$, and $[OH^-]$ of this hand soap. Is this hand soap an acidic or basic substance? Is this hand soap an electrophile or nucleophile?

3. Which of these common substances would you expect to be acidic?
Pepsi®, clothes washing detergent, glass cleaner, wine, dish washing detergent

4. Rainwater has a pH of ~5.7. Suggest why rainwater is not a neutral mixture. Hint: This is a normal process that has nothing to do with pollution.

Experiment 19 Lab Report (Show ALL Work and ALWAYS include units)

Part A – Determine pH of Several Acid Solutions

1. Measured pH of 1.0 M $HCl(aq)$ ________

2. Measured pH of 0.50 M $HCl(aq)$ ________

3. Measured pH of 0.25 M $HCl(aq)$ ________

4. Measured pH of 0.10 M $HCl(aq)$ ________

5. Measured pH of 1.0 M $CH_3COOH(aq)$ ________

6. Measured pH of 0.50 M $CH_3COOH(aq)$ ________

7. Measured pH of 0.25 M $CH_3COOH(aq)$ ________

8. Measured pH of 0.10 M $CH_3COOH(aq)$ ________

Part B – Determine pH of Several Base Solutions

9. Measured pH of 1.0 M $NaOH(aq)$ ________

10. Measured pH of 0.50 M $NaOH(aq)$ ________

11. Measured pH of 0.25 M $NaOH(aq)$ ________

12. Measured pH of 0.10 M $NaOH(aq)$ ________

13. Measured pH of 1.0 M $NH_3(aq)$ ________

14. Measured pH of 0.50 M $NH_3(aq)$ ________

15. Measured pH of 0.25 M $NH_3(aq)$ ________

16. Measured pH of 0.10 M $NH_3(aq)$ ________

Part C – Determine pH of Several Common Household Solutions

17. Measured pH of lemon juice ________

18. Measured pH of Liquid Plumber® ________

19. Measured pH of Windex® ________

20. Measured pH of Coca Cola® ________

21. Measured pH of vinegar ________

22. Measured pH of tap water ________

23. Measured pH of distilled or deionized water ________

24. What happens to the pH of both $HCl(aq)$ and $CH_3COOH(aq)$ as the concentration changes from 1.0 *M* to 0.50 *M* to 0.25 *M* to 0.10 *M*? Explain why this happens.

25. Is there a noticeable difference between the pH of $HCl(aq)$ and $CH_3COOH(aq)$? Is there a noticeable difference between the pH of $NaOH(aq)$ and $NH_3(aq)$? If there is a difference, explain why. If there is not a difference, explain why.

26. Is there a trend in the pH of the two cleaning solutions Liquid Plumber® and Windex®?

27. Is there a trend in the pH of the food substances vinegar, Coca Cola®, and lemon juice?

28. Based upon your answers to questions 26 and 27, suggest what types of foods humans prefer to consume.

29. Compare the pH of tap water and distilled or deionized water? Is there a difference in the pH of the two? If there is a difference, explain why. If there is not a difference, explain why not.

Experiment 20
Enthalpy, Entropy, and Gibbs Free Energy - Is That Reaction Spontaneous or Not?

Objectives

- Understand the contributions of enthalpy and entropy in Gibbs free energy.
- Understand what spontaneous means.
- Understand how spontaneity can be affected by temperature.
- Realize that thermodynamics and kinetics interact with one another to make a reaction occur in a suitable time frame.

Topics Introduced in this Lab

- Enthalpy
- Entropy
- Gibbs free energy
- Reaction spontaneity

Related Zumdahl Sections

- Chapter 16 Sections 1, 3, 4, and 5

Safety Considerations

- 30% hydrogen peroxide is a strong oxidizing agent. Contact with the skin or eyes can cause rapid and potentially painful oxidation. If any hydrogen peroxide comes into contact with your skin or eyes, wash with copious amounts of water.
- Saturated KI(*aq*) is toxic if swallowed.
- Do not ingest any of these compounds.
- Wear safety glasses and appropriate lab wear at all times while in the lab.

Introduction

J. Willard Gibbs was the first scientist to recognize the importance of enthalpy and entropy in predicting whether a chemical reaction or process will occur. This is expressed in **Equation 1**. *If the ΔG value is negative, the reaction is spontaneous. Positive values of ΔG indicate a nonspontaneous reaction. A ΔG value of 0 indicates the process or reaction is at equilibrium.* Spontaneous reactions or processes will occur if given sufficient time.

$$\Delta G = \Delta H - T\Delta S \text{ at constant pressure}$$

where ΔG = change in Gibbs free energy (kJ/mol)
ΔH = change in Enthalpy (kJ/mol)
T = temperature (K)
ΔS = change in entropy (kJ/mol×K)

Equation 1. Gibbs Free Energy Relationship

Since both ΔH and ΔS can be either positive or negative, there are four ΔG possibilities that can result, as indicated in **Table 1**.

ΔH	ΔS	ΔG	Possibility
> 0	< 0	> 0 at all temperatures	1
> 0	> 0	< 0 at high temperatures > 0 at low temperatures	2
< 0	< 0	> 0 at high temperatures < 0 at low temperatures	3
< 0	> 0	< 0 at all temperatures	4

Table 1. Gibbs Free Energy Possibilities Based Upon Enthalpy and Entropy Values

Let's look at the four possibilities. Possibility 1, ΔH positive (an increase in enthalpy) and ΔS negative (a decrease in entropy), yields only positive ΔG values no matter the temperature. In other words, all reactions and processes having positive ΔH and negative ΔS are nonspontaneous. Possibility 4 is spontaneous at all temperatures because it has negative ΔH and positive ΔS. Possibilities 2 and 3 are temperature dependent. As the $T\Delta S$ term in **Equation 1** becomes larger or smaller, the reaction or process becomes spontaneous or nonspontaneous. Possibility 3, negative ΔH and ΔS, is spontaneous at low temperatures (when the $T\Delta S$ term is small relative to ΔH), and nonspontaneous at high temperatures. Possibility 2, positive ΔH and ΔS, is spontaneous at high temperatures (when the $T\Delta S$ term is large relative to ΔH), and nonspontaneous at low temperatures.

Values of $\Delta H^{\circ}_{\text{reaction}}$ are determined by taking the sum of the reaction stoichiometric coefficients multiplied by the $\Delta H^{\circ}_{\text{f}}$ values for the products, then subtracting the sum of the reaction stoichiometric coefficients multiplied by the $\Delta H^{\circ}_{\text{f}}$ values for the reactants, as shown in **Equation 2**. $\Delta S^{\circ}_{\text{reaction}}$ is calculated similarly using stoichiometric coefficient values and S° values for the products and reactants as indicated in **Equation 3**.

$$\Delta H^{\circ}_{\text{reaction}} = \sum n_{\text{products}} \Delta H^{\circ}_{\text{f, products}} - \sum n_{\text{reactants}} \Delta H^{\circ}_{\text{f, reactants}}$$

Equation 2. Calculation of $\Delta H^{\circ}_{\text{reaction}}$

$$\Delta S^{\circ}_{\text{reaction}} = \sum n_{\text{products}} S^{\circ}_{\text{products}} - \sum n_{\text{reactants}} S^{\circ}_{\text{reactants}}$$

Equation 3. Calculation of $\Delta S^{\circ}_{\text{reaction}}$

In this experiment you will calculate the $\Delta H^{\circ}_{\text{reaction}}$ and $\Delta S^{\circ}_{\text{reaction}}$ values for several reactions. Based upon those calculated values, you can determine which of the four

ΔG possibilities is appropriate for each reaction. Then you will examine a reaction to see how thermodynamics and kinetics interact.

Procedure

Part A – Calculation of $\Delta H^\circ_{\text{reaction}}$ and $\Delta S^\circ_{\text{reaction}}$ Values for Several Reactions

1. Given below in **Table 2** are the ΔH°_f and S° values for several substances.

Substance	ΔH°_f (kJ/mol)	S° (kJ/mol K)
Al(*s*)	0.0	0.028
$AlCl_3$(*s*)	-704	0.111
C(*s*, graphite)	0.0	0.006
Cl_2(*g*)	0.0	0.223
CO(*g*)	-110.5	0.198
CO_2(*g*)	-393.5	0.214
F_2(*g*)	0.0	0.203
H_2(*g*)	0.0	0.131
H_2O(*g*)	-286	0.070
H_2O_2(*l*)	-191.2	0.144
N_2(*g*)	0.0	0.192
N_2F_4(*g*)	-8.4	0.301
N_2O(*g*)	82	0.220
NCl_3(*g*)	230	0.293
NF_2(*g*)	42.3	0.250
NF_3(*g*)	-132.1	0.261
NH_3(*g*)	-46	0.211
NO(*g*)	90	0.211
NOCl(*g*)	51.7	0.262
O_2(*g*)	0.0	0.205
P(*s*)	0.0	0.041
PH_3(*g*)	22.9	0.210

Table 2. Selected Thermodynamic Data

Use these values to calculate $\Delta H^\circ_{\text{reaction}}$ and $\Delta S^\circ_{\text{reaction}}$ for the following reactions.

$N_2(g) + 3F_2(g) \rightarrow 2NF_3(g)$
$N_2(g) + 3Cl_2(g) \rightarrow 2NCl_3(g)$
$N_2F_4(g) \rightarrow 2NF_2(g)$
$2N_2O(g) \rightarrow 2N_2(g) + O_2(g)$
$C(s, \text{graphite}) + O_2(g) \rightarrow CO_2(g)$

$2Al(s) + 3Cl_2(g) \rightarrow 2AlCl_3(s)$
$2NOCl(g) \rightarrow 2NO(g) + Cl_2(g)$
$4NO(g) + 6H_2O(g) \rightarrow 4NH_3(g) + 5O_2(g)$

2. Based upon the calculated $\Delta H^{\circ}_{\text{reaction}}$ and $\Delta S^{\circ}_{\text{reaction}}$ values, determine which of the four ΔG possibilities is appropriate for each reaction.

Part B – Interaction of Thermodynamics and Kinetics

3. Obtain one or two wooden splints, 5 mL of 30% H_2O_2, and 1-2 mL of saturated KI(aq).
4. Using data from **Table 2** calculate the $\Delta H^{\circ}_{\text{reaction}}$ and $\Delta S^{\circ}_{\text{reaction}}$ for this reaction:

$$2\,H_2O_2 \rightarrow O_2 + 2H_2O$$

5. This reaction is an example of ΔG possibility 4, spontaneous at all temperatures. Since this decomposition reaction produces oxygen, O_2, the presence of oxygen is a good indicator of the extent of reaction. An easy test for O_2 production uses a glowing wooden splint. When placed into O_2 a glowing wooden splint immediately ignites.
6. Place 2-3 mL of 30% H_2O_2 into a large test tube. Securely attach the test tube to the base of a ring stand using a utility clamp, **Figure 1**. This will prevent you from knocking over the test tube and spilling the hydrogen peroxide.

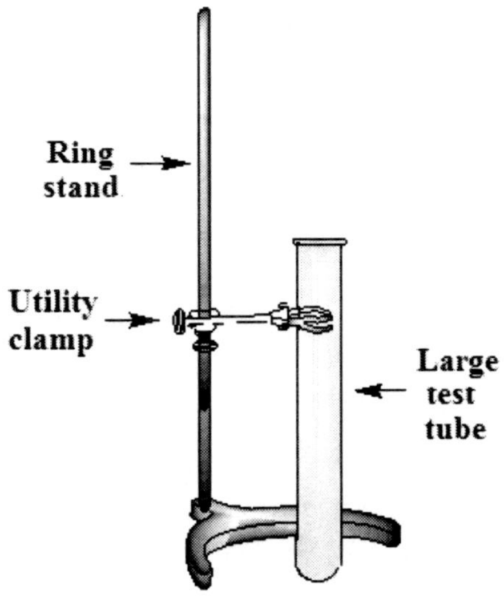

Figure 1. Experimental Setup

7. Using lab tongs, hold on to one end of a wooden splint. Using a match, lighter, or Bunsen burner, ignite the other end of the wooden splint. Once the splint ignites, blow it out, leaving glowing embers. Holding the splint with the tongs, insert the glowing end of the splint into the mouth of the test tube but do not insert the splint into the hydrogen peroxide.

8. Observe and record what happens to the splint after it is inserted into the test tube.
9. To avoid hydrogen peroxide contamination, pour the contents of the test tube into the waste disposal bottle provided by your instructor. Reattach the test tube to the ring stand. Pour a fresh 2-3 mL of 30% H_2O_2 into the test tube, then continue the experiment.
10. A catalyst for the decomposition of H_2O_2 is KI. Using a medicine dropper, drop one or two drops of saturated KI(*aq*) solution into the 30% H_2O_2. **Do not stand too close to the test tube nor look straight down into the test tube.**
11. Observe and record what happens when the KI is added to the 30% H_2O_2.
12. Quickly repeat steps 7 and 8.

Experiment 20 Pre-Lab (Show ALL Work and ALWAYS include units)

1. Using data from **Table 2**, calculate the $\Delta H^{\circ}_{\text{reaction}}$ for this reaction:

$$2PH_3(g) \rightarrow 2P(s) + 3H_2(g)$$

2. Using data from **Table 2**, calculate the $\Delta S^{0}_{\text{reaction}}$ for this reaction:

$$2PH_3(g) \rightarrow 2P(s) + 3H_2(g)$$

3. This reaction is an example of which one of the ΔG possibilities?

4. Calculate the absolute temperature at which this reaction becomes spontaneous.

Experiment 20 Lab-Report (Show ALL Work and ALWAYS include units)

Part A – Calculation of $\Delta H^{o}_{reaction}$ and $\Delta S^{o}_{reaction}$ Values for Several Reactions

Show ALL work on separate sheet(s) of paper and turn in your work with the lab report. Record your answers in the table below.

Reaction	$\Delta H^{o}_{reaction}$ (kJ/mol)	$\Delta S^{o}_{reaction}$ (kJ/mol K)	ΔG Possibility
$N_2 + 3F_2 \rightarrow 2NF_3$			
$N_2 + 3Cl_2 \rightarrow 2NCl_3$			
$N_2F_4 \rightarrow 2NF_2$			
$2N_2O \rightarrow 2N_2 + O_2$			
$C + O_2 \rightarrow CO_2$			
$2Al + 3Cl_2 \rightarrow 2AlCl_3$			
$2NOCl \rightarrow 2NO + Cl_2$			
$4NO + 6H_2O \rightarrow 4NH_3 + 5O_2$			

Part B – Interaction of Thermodynamics and Kinetics

1. Describe what happened when you placed the burning splint into the test tube containing 30% H_2O_2.

2. Describe what happened when you dropped the KI(*aq*) into the test tube containing 30% H_2O_2.

3. Describe what happened when you placed the burning splint into the test tube containing 30% H_2O_2 after the KI(*aq*) was added.

4. For all reactions in Part A classified as either ΔG possibilities 2 or 3, determine the absolute temperature at which the reaction becomes spontaneous. Show all work.

5. Explain your results from Part B when you placed the burning splint into the test tube containing solely 30% H_2O_2, expressly addressing the fact that H_2O_2 decomposition is spontaneous at all temperatures.

6. Explain your results from Part B when you placed the burning splint into the test tube containing 30% H_2O_2 and a drop or two of saturated KI(*aq*), addressing the fact that H_2O_2 decomposition is spontaneous at all temperatures.

7. Other catalysts for the decomposition of H_2O_2 include MnO_2, PbO_2, Fe_2O_3, and light. Use this information to explain why the bottle of H_2O_2 in your bathroom cabinet is brown-colored not opaque.

Experiment 21
Galvanic Cells –
Making Electrochemical Batteries

Objectives

- Understand how simple batteries are made.
- Understand that battery voltage is a function of the chemicals used to make the battery.

Topics Introduced in this Lab

- Galvanic cells
- Standard cells
- Nonstandard cells
- Cell potentials (voltages)

Related Zumdahl Sections

- Chapter 17 Sections 2, 3, 4, 5, and 6

Safety Considerations

- Do not allow the solutions to come into contact with your skin or eyes. All of these solutions contain heavy metal salts that are toxic to humans. Dispose of solutions in an appropriate waste container.
- Lead (Pb) metal and salts should be carefully handled. Wash your hands immediately before leaving lab and prior to eating.
- Do not ingest any of these compounds.
- Wear safety glasses and appropriate lab wear at all times while in the lab.

Introduction

Luigi Galvani and Alessandro Volta were among the first scientists to explore generating electricity with batteries. The first batteries were very crude. Galvani's version of a battery consisted of metallic copper plates alternating with zinc plates. Each plate was separated from the other with a piece of cloth soaked in salt water. The voltmeter was even more crude. Galvani used the legs of a dead frog hooked to wires. When electricity was applied to the frog legs, they jumped and wriggled enough that Galvani could use the motion to quantify the power of the battery. Technically, batteries are called Galvanic cells to honor Galvani. Electrical potential difference is called the volt to honor Volta, who also worked on batteries.

Today we would make Galvani's battery as shown in **Figure 1**. It consists of a strip of Cu metal dipped into $Cu(NO_3)_2(aq)$ and a strip of Zn metal dipped into $Zn(NO_3)_2(aq)$. The two metal strips are connected to wires, the other ends of which are connected to an electronic voltmeter. Completing the battery is a salt bridge consisting of a glass tube filled with a paste made of NaCl and just enough water to make the paste sticky. Typically, a plug of fiberglass wool is stuck into the ends of the salt bridge to keep the NaCl from falling out of the bridge. The purpose of the salt bridge is to allow Na^+ and Cl^- ions to flow into the two aqueous solutions to maintain electrical neutrality in each beaker.

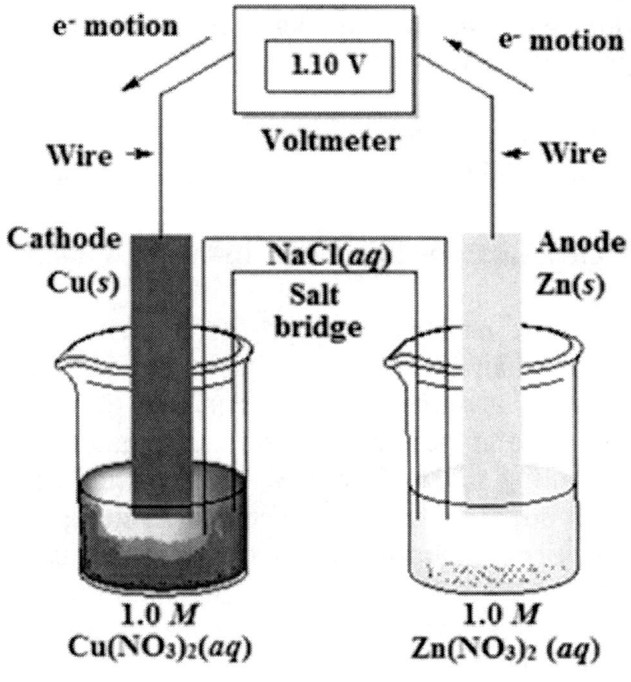

Figure 1. Cu-Zn Galvanic Cell Using Beakers

Disposal of heavy metal solutions is problematic. Miniaturization of this experiment makes the experiment more eco-friendly. To do this, we replace the beakers and their 50 or more mL of solution with a few drops of solution in a well plate, **Figure 2**. A salt bridge can be constructed of filter paper strips soaked in NaCl(*aq*). Small metal strips are attached to the voltmeter using alligator clips and wire.

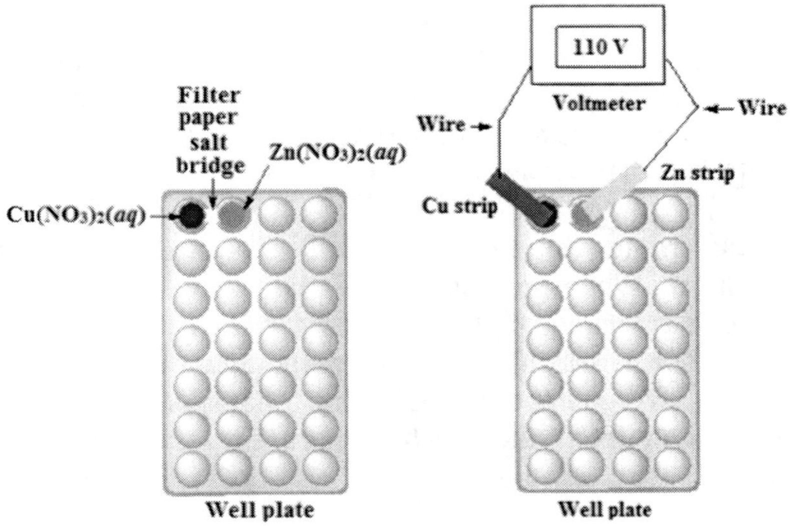

Figure 2. Cu-Zn Galvanic Cell Using a Well Plate

As the battery operates, an oxidation-reduction (redox) reaction occurs. Oxidation, loss of electrons, occurs at one metal strip. Simultaneously, reduction, gaining of electrons, occurs at the other metal strip. Electrons produced and consumed in the redox reactions pass through the wire and voltmeter to complete the electrical circuit, **Equation 1**.

$$Cu^{2+} + 2e^- \rightarrow Cu \quad \textbf{reduction half-reaction}$$
$$Zn \rightarrow Zn^{2+} + 2e^- \quad \textbf{oxidation half-reaction}$$

Equation 1. Redox Half-Reactions Occurring in Cu-Zn Galvanic Cell

In the Cu-Zn galvanic cell, the copper metal strip is the cathode, where reduction occurs. In this half-cell Cu^{2+} ions from the $Cu(NO_3)_2(aq)$ solution are attracted to the metal strip, acquire two electrons, and then plate out onto the Cu strip. When the circuit is complete, the Cu metal strip gains mass as $Cu^{2+}(aq)$ ions react from solution, forming more Cu metal. At the anode (the Zn metal strip), oxidation occurs as Zn metal loses two electrons forming $Zn^{2+}(aq)$ ions, which dissolve into the $Zn(NO_3)_2(aq)$ solution. As the cell operates, the Zn metal strip loses mass as Zn becomes Zn^{2+} ions that dissolve into the solution. Without a salt bridge the Cu half-cell side would become negatively charged as Cu^{2+} ions are removed from solution. Simultaneously, Zn^{2+} ions flowing into the Zn oxidation half-cell side would make the $Zn(NO_3)_2(aq)$ positive. However, if a NaCl salt bridge is present, $Na^+(aq)$ ions move into the Cu side while $Cl^-(aq)$ ions move into the Zn side, maintaining cell neutrality.

As long as cell conditions are constant, each cell combination produces a well-known initial voltage. To maintain consistency, scientists have defined a standard set of conditions to measure cell voltages. Those conditions are 25.0 °C, 1.00 atm of pressure, metals in their solid state, and 1.00 *M* solution concentrations. In most college chemistry labs the temperature, pressure, and physical states of the metals are essentially at standard conditions. To make the conditions nonstandard, changing solution concentrations is easiest.

In this experiment you will make a series of galvanic cells using Fe metal as one electrode and Ni, Cu, Zn, Al, Mg, or Pb metal strips as the other electrode. Each strip will be dipped into a solution containing ions of that metal. You will use aqueous solutions of $Fe(NO_3)_2$, $Ni(NO_3)_2$, $Cu(NO_3)_2$, $Zn(NO_3)_2$, $Al(NO_3)_3$, $Mg(NO_3)_2$, and $Pb(NO_3)_2$.

Using an electronic voltmeter, you will measure the electric potential generated by each cell. When the positive voltmeter lead is attached to the anode and negative lead is attached to the cathode, a positive voltage appears on the voltmeter because the reaction is occurring spontaneously. If the leads are reversed, a negative voltage appears, indicating a nonspontaneous reaction. From the voltage sign you can determine which metals are the cathode and anode in each galvanic cell.

Procedure

Part A – Assemble the Galvanic Cells

1. Obtain one well plate and ~ 10 mL of 1.00 $M(aq)$ $Fe(NO_3)_2$ and ~ 2 mL each of 1.00 $M(aq)$ $Ni(NO_3)_2$, $Cu(NO_3)_2$, $Zn(NO_3)_2$, $Al(NO_3)_3$, $Mg(NO_3)_2$, and $Pb(NO_3)_2$ solutions.
2. Using tape and a grease pencil, label each well in your well plate as indicated in **Figure 3**.
3. Use a clean medicine dropper or plastic pipet to fill the six wells in one row of the well plate with 1.00 M $Fe(NO_3)_2(aq)$, as shown in **Figure 3**.
4. Use a clean medicine dropper or plastic pipet to fill six wells in the well plate row adjacent to the one containing 1.00 M $Fe(NO_3)_2(aq)$ with 1.00 $M(aq)$ $Ni(NO_3)_2$, $Cu(NO_3)_2$, $Zn(NO_3)_2$, $Al(NO_3)_3$, $Mg(NO_3)_2$, and $Pb(NO_3)_2$ solutions as shown in **Figure 3**. In between filling each well be certain to wash the medicine dropper or pipet with distilled or deionized water so that the solutions are not contaminated with residue from the previous solution. (Remember that all waste from this experiment must go into the appropriate waste container.)

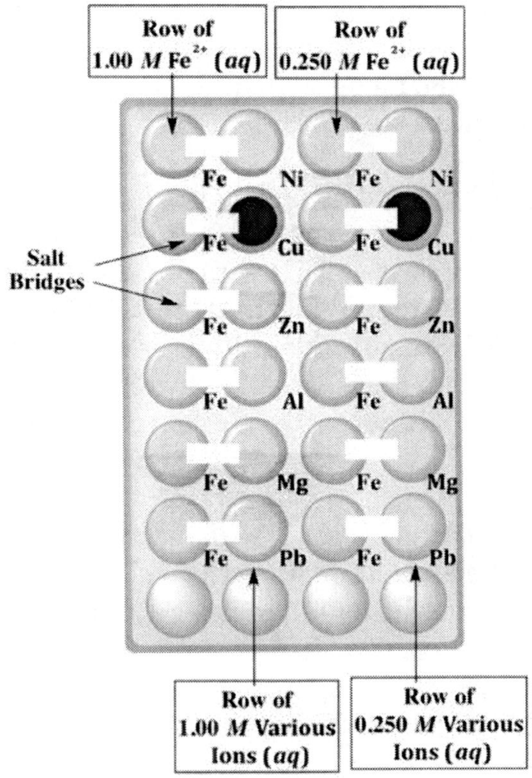

Figure 3. Well Plate Configuration Experimental Setup

5. Obtain ~ 10 mL of 0.250 $M(aq)$ $Fe(NO_3)_2$ and ~ 2 mL each of 0.250 $M(aq)$ $Ni(NO_3)_2$, $Cu(NO_3)_2$, $Zn(NO_3)_2$, $Al(NO_3)_3$, $Mg(NO_3)_2$, and $Pb(NO_3)_2$ solutions.

6. Repeat steps 3 and 4 to fill the 3rd and 4th rows of the well plate with 0.250 *M* $Fe(NO_3)_2(aq)$ and 0.250 *M*(*aq*) solutions of $Ni(NO_3)_2$, $Cu(NO_3)_2$, $Zn(NO_3)_2$, $Al(NO_3)_3$, $Mg(NO_3)_2$, and $Pb(NO_3)_2$ as shown in **Figure 3**.
7. Using tweezers place salt bridges made of filter paper strips soaked in NaCl(*aq*) to bridge solutions of the same concentration as shown in **Figure 3**.
8. Obtain small metal strips of Fe, Ni, Cu, Zn, Al, Mg, and Pb, as well as an electronic voltmeter, wires, and attached alligator clips.

<u>Part B – Measure the Voltage Produced by Galvanic Cells Containing 1.00 *M* Solutions</u>

9. Attach the positive voltmeter wire to the Fe metal strip and place it into the 1.00 *M* $Fe(NO_3)_2(aq)$ adjacent to the Ni solution. Attach the negative voltmeter wire to the Ni metal strip and place it into the 1.00 *M* $Ni(NO_3)_2(aq)$. Read the measured voltage indicated on the voltmeter. Be sure to record the measured voltage and the polarity (positive or negative).
10. Remove the Fe metal strip and wash it with distilled or deionized water from a wash bottle. Place the Fe metal strip in the well adjacent to the 1.00 *M* (*aq*) $Cu(NO_3)_2$. Remove the strip of Ni metal, rinse it with distilled or deionized water, and replace it with the Cu strip. Place the Cu strip into the 1.00 *M* $Cu(NO_3)_2(aq)$. Read the measured voltage indicated on the voltmeter. Be sure to record the measured voltage and the polarity (positive or negative).
11. Repeat step 10 using the Fe metal strip in conjunction with the Zn, Al, Mg, and Pb strips in their associated solutions.

<u>Part C – Measure the Voltage Produced by Galvanic Cells Containing 0.250 *M* Solutions</u>

12. Remove the Fe metal strip and wash it with distilled or deionized water from a wash bottle. Place the Fe metal strip into the 0.250 *M* $Fe(NO_3)_2(aq)$ adjacent to the Ni solution. Attach the negative voltmeter wire to the freshly cleaned Ni metal strip and place it into the 0.250 *M* $Ni(NO_3)_2(aq)$. Read the measured voltage indicated on the voltmeter. Be sure to record the measured voltage as well as the polarity (positive or negative).
13. Remove the Fe metal strip and wash it with distilled or deionized water from a wash bottle. Place the Fe metal strip in the well adjacent to the 0.250 *M* $Cu(NO_3)_2(aq)$. Remove the strip of Ni metal, rinse it with distilled or deionized water, and replace it with the freshly cleaned Cu strip. Place the Cu strip into the 0.250 *M* $Cu(NO_3)_2(aq)$. Read the measured voltage indicated on the voltmeter. Be sure to record the measured voltage as well as the polarity (positive or negative).
14. Repeat step 13 using the Fe metal strip in conjunction with the Zn, Al, Mg, and Pb strips in their associated solutions.

<u>Part D – Measure the Voltage Produced by Galvanic Cells Containing 1.00 *M* and 0.250 *M* Solutions</u>

15. Using tweezers, carefully remove all of the NaCl(*aq*) salt bridges. Be careful not to contaminate any of the solutions as you remove the salt bridges. Now

place new salt bridges so that they bridge a well containing 0.250 *M* $Fe(NO_3)_2(aq)$ and 1.00 *M*(aq) solutions of each of the other solutions. In other words, salt bridges will connect the two center rows of the well plate shown in **Figure 3**.
(All used salt bridges should be disposed of in the appropriate container.)

16. Remove the Fe metal strip and wash it with distilled or deionized water from a wash bottle. Place the Fe metal strip into the 0.250 *M* $Fe(NO_3)_2(aq)$ adjacent to the Ni solution. Attach the negative voltmeter wire to the freshly cleaned Ni metal strip and place it into the 1.00 *M* $Ni(NO_3)_2(aq)$. Read the measured voltage indicated on the voltmeter. Be sure to record the measured voltage as well as the polarity
(positive or negative).
17. Remove the Fe metal strip and wash it with distilled or deionized water from a wash bottle. Place the Fe metal strip in the well adjacent to the 1.00 *M*(aq) $Cu(NO_3)_2$. Remove the strip of Ni metal, rinse it with distilled or deionized water, and replace it with the freshly cleaned Cu strip. Place the Cu strip into the 1.00 *M* $Cu(NO_3)_2(aq)$. Read the measured voltage indicated on the voltmeter. Be sure to record the measured voltage as well as the polarity (positive or negative).
18. Repeat step 17 using the Fe metal strip in conjunction with the Zn, Al, Mg, and Pb strips in their associated solutions.

Experiment 21 Pre-Lab (Show ALL Work and ALWAYS include units)

1. A galvanic cell is made from a Ca metal strip dipped into 1.00 *M* $Ca(NO_3)_2(aq)$ and a Sn metal strip dipped into 1.00 *M* $Sn(NO_3)_2(aq)$. A salt bridge made of $NaCl(aq)$ paste connects the two cells. The positive voltmeter wire is attached to the Ca strip while the negative voltmeter wire is attached to the Sn strip. Measured voltage for this galvanic cell is **–2.62 V**. In this cell, which metal is the cathode and which is the anode?

2. For the Ca-Sn galvanic cell described in question 1, write the half-reactions for each cell.

3. For the Ca-Sn galvanic cell described in question 1, which metal strip will gain mass and which will lose mass?

4. Carefully and completely explain how the metal strips in the Ca-Sn galvanic cell gain or lose mass.

5. Carefully and completely explain what role the salt bridge plays in the Ca-Sn galvanic cell. What ions flow from the bridge into each half-cell and why?

Experiment 21 Lab Report (Show ALL Work and ALWAYS include units)

Part B – Measure the Voltage Produced by Galvanic Cells Containing 1.00 *M* Solutions

1. Measured Sign and Voltage of Fe-Ni Galvanic cell ____________

2. Measured Sign and Voltage of Fe-Cu Galvanic cell ____________

3. Measured Sign and Voltage of Fe-Zn Galvanic cell ____________

4. Measured Sign and Voltage of Fe-Al Galvanic cell ____________

5. Measured Sign and Voltage of Fe-Mg Galvanic cell ____________

6. Measured Sign and Voltage of Fe-Pb Galvanic cell ____________

Part C – Measure the Voltage Produced by Galvanic Cells Containing 0.250 *M* Solutions

7. Measured Sign and Voltage of Fe-Ni Galvanic cell ____________

8. Measured Sign and Voltage of Fe-Cu Galvanic cell ____________

9. Measured Sign and Voltage of Fe-Zn Galvanic cell ____________

10. Measured Sign and Voltage of Fe-Al Galvanic cell ____________

11. Measured Sign and Voltage of Fe-Mg Galvanic cell ____________

12. Measured Sign and Voltage of Fe-Pb Galvanic cell ____________

Part D – Measure the Voltage Produced by Galvanic Cells Containing 1.00 *M* and 0.250 *M* Solutions

13. Measured Sign and Voltage of Fe-Ni Galvanic cell ____________

14. Measured Sign and Voltage of Fe-Cu Galvanic cell ____________

15. Measured Sign and Voltage of Fe-Zn Galvanic cell ____________

16. Measured Sign and Voltage of Fe-Al Galvanic cell ____________

17. Measured Sign and Voltage of Fe-Mg Galvanic cell ____________

18. Measured Sign and Voltage of Fe-Pb Galvanic cell ____________

19. From the measured voltage for each cell determine which metal is the anode and cathode in each galvanic cell.

	Anode Metal	Cathode Metal
Fe-Ni galvanic cell	____________	____________
Fe-Cu galvanic cell	____________	____________
Fe-Zn galvanic cell	____________	____________
Fe-Al galvanic cell	____________	____________
Fe-Mg galvanic cell	____________	____________
Fe-Pb galvanic cell	____________	____________

20. Which metal strip loses mass and gains mass in each galvanic cell?

	Metal Loses Mass	Metal Gains Mass
Fe-Ni galvanic cell	____________	____________
Fe-Cu galvanic cell	____________	____________
Fe-Zn galvanic cell	____________	____________
Fe-Al galvanic cell	____________	____________
Fe-Mg galvanic cell	____________	____________
Fe-Pb galvanic cell	____________	____________

21. Write redox half-reactions for the Fe-Cu and Fe-Zn galvanic cells.

Fe-Cu galvanic cell **Fe-Zn galvanic cell**

22. How did the cell voltage for each cell in Part B compare to that cell's voltage in Part C?

23. How did the cell voltage for each cell in Part B compare to that cell's voltage in Part D?

24. Cell voltages in standard and nonstandard cells are described by the Nernst equation. Explain your results from Lab Report question 22 based upon the Nernst equation.

$$\text{Nernst Equation} \quad \varepsilon = \varepsilon^{o} - \frac{RT}{nF}\log Q$$

where ε = nonstandard cell voltage, ε^{o} = standard cell voltage, R = universal gas constant, T = Kelvin temperature, n = number of electrons involved in redox reaction, F = Faraday constant, and Q = reaction quotient.

25. Explain your results from Lab Report question 23 based upon the Nernst equation.

Experiment 22
Exploring Electrolysis

Objective

- Operate an electrolytic cell, then use your knowledge of electrolysis along with the ideal gas law to identify an unknown metal.

Topics Introduced in this Lab

- Electrolysis

Related Zumdahl Sections

- Chapter 17 Section 8

Safety Considerations

- Wear safety glasses and appropriate lab wear at all times while in the lab.
- Avoid getting sodium sulfate or acetic acid solution on your skin or eyes. If any does contact your skin or eyes, rinse the area immediately with water followed by washing with soap and water.
- Some of the solutions produced in this experiment by oxidation of the metal electrodes are toxic and should be handled with care and disposed of appropriately. Some are also corrosive. If you get any on your skin or eyes, rinse immediately with water and alert your instructor.
- The reaction in this experiment produces H_2 gas. H_2 gas is highly flammable. No open flames should be present in the lab during this experiment.
- Do not ingest any of the chemicals in this experiment.

Introduction

Galvanic cells use two "isolated" redox half-reactions to convert chemical energy into electrical energy. Separating the oxidation and reduction halves of the redox reaction creates a potential difference between one half-reaction and the other. If a medium which conducts electrons, such as an electrical wire, connects the two half-reactions, the potential difference causes electrons to flow, creating electrical current. A typical battery is an example of a galvanic cell.

Electrolytic cells, on the other hand, convert electrical energy into chemical reactions. Electrical voltage drives electrons from one half-reaction to the other in a non-spontaneous direction. This process is called electrolysis.

In this experiment, you will build and operate an electrolytic cell, **Figure 1**. One cell electrode will be an unknown metal. The other electrode will be copper wire. The unknown metal will be the anode where the metal is oxidized, forming its cation in solution. As the electrolytic cell operates, the metal anode's mass will slowly decrease. Copper wire is the cathode where H^+ in solution is reduced to H_2 gas. The cell solution is 0.50 *M* sodium sulfate / 0.50 *M* acetic acid. Acetic acid is the source of H^+ that is reduced to H_2 gas.

Any anode you will use in this experiment is a metal that readily oxidizes to form 2+ cations. Two electrons are lost for each metal atom and, therefore, two moles of electrons are lost for each mole of oxidized metal. Thus, if the total moles of electrons lost during operation of the electrolytic cell is determined, the moles of metal oxidized can also be determined. Measuring the mass change of the metal during the reaction permits determination of the molar mass of the metal (*M*) and its identity, by using **Equation 1**.

$$\mathbf{M_{(s)} \rightarrow M^{2+}{}_{(aq)} + 2e^-\ ;\ therefore\ moles\ e^-\left(\frac{1\ mole\ \mathit{M}}{2\ moles\ e^-}\right) = moles\ \mathit{M}}$$

$$\mathbf{\frac{grams\ \mathit{M}}{moles\ \mathit{M}} = Molar\ Mass\ of\ Metal}$$

Equation 1. Determining Molar Mass from Moles of e⁻ and Mass Loss of Metal

To determine the moles of electrons lost and hence the moles of metal oxidized during the course of the reaction, we will use the ideal gas law. Recall that H^+ is being reduced to H_2 gas while the metal is being oxidized. This half-reaction is represented by **Equation 2**.

$$\mathbf{2H^+(\mathit{aq}) + 2e^- \rightarrow H_2(\mathit{g})}$$

Equation 2. Reduction of H^+ to H_2 Gas

For every mole of H_2 produced, two moles of electrons are used. If the moles of H_2 gas produced are measured, the moles of electrons consumed in the reaction can be calculated. (It is important to remember that these are the same electrons emitted by the metal oxidation.) To calculate the moles of H_2, we will determine the volume, temperature, and pressure of the H_2 gas produced in the reaction. Then, we can calculate the moles of H_2 produced using **Equation 3**.

$$n = \frac{PV}{RT}$$

Equation 3. Calculation of the Moles of Gas from the Ideal Gas Law

Procedure

1. Fill a 100-mL graduated cylinder to the top with the 0.50 *M* sodium sulfate / 0.50 *M* acetic acid solution. Place a 600-mL beaker upside down on the top of the graduated cylinder, **Figure 1a**. Holding the beaker and graduated cylinder together, flip the assembly so the beaker is now upright and the graduated cylinder is upside down. Immediately pour approximately 300 mL of additional solution into the beaker. You should be able to lift the graduated cylinder slightly off the bottom of the beaker without the solution level in the cylinder decreasing significantly. (You want as little air as possible above the solution column). Place

the beaker/graduated cylinder assembly on the bench top and secure the graduated cylinder approximately 1 or 2 inches from the bottom of the beaker using a ring stand and utility clamp, **Figure 1b**.

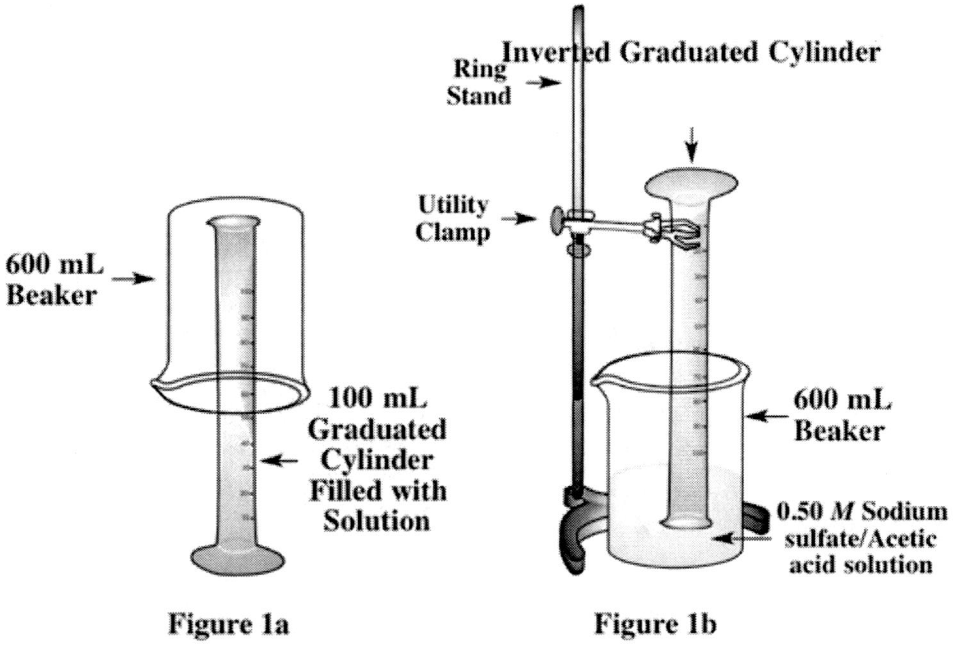

Figure 1. Setting up the Gas Collection Apparatus

2. Using wire strippers, remove the casing from roughly 1 inch of one end of a 10-inch-long piece of copper wire and roughly 2 inches from the other end. Both ends of the copper wire should be exposed.
3. Coil the 2-inch exposed end of copper wire around a pencil or pen creating a permanent coil in the wire. Bend the wire below the coil to create a "J" shape. Bend the other end of the wire below the exposed portion so that the wire hangs over the edge of the beaker, **Figure 2a**.
4. Position the wire in the beaker so that the coiled portion is submerged and completely inside the inverted graduated cylinder, **Figure 2b**. You may need to lower the graduated cylinder closer to the beaker bottom to ensure that all exposed copper on the coiled end of the wire is inside the graduated cylinder. The other end of the wire must remain over and outside the edge of the beaker.

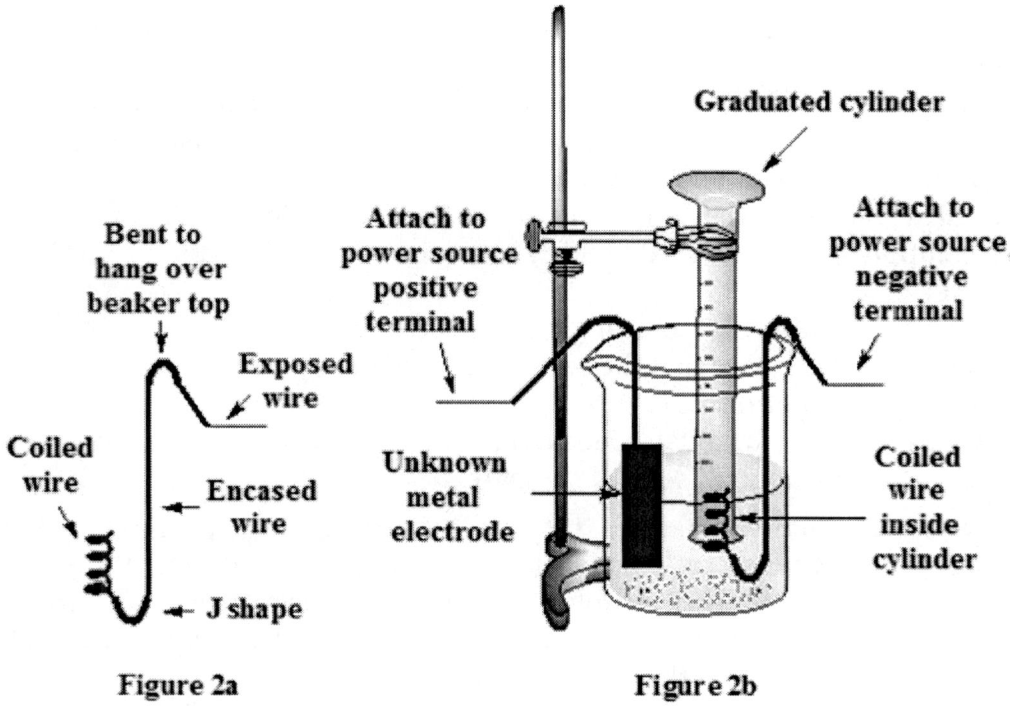

Figure 2. Coiled Wire and Experimental Setup

5. Obtain an unknown metal electrode. Clean the surface of the electrode using a piece of sand paper.
6. Rinse off your electrode with water, ensuring that any rinse waste goes into the waste container. Clean and dry the electrode thoroughly.
7. Record the mass of the clean unknown metal electrode as well as its number.
8. Place the unknown metal electrode inside the beaker, leaning it against the side with the bottom portion of the electrode submerged in solution, **Figure 2b**.
9. Ask your instructor whether you will be using a wall or battery power source. Obtain whichever power source you will use along with two alligator clip wires.
10. Attach one alligator clip of one wire to the top of the unknown metal electrode. The alligator clip must remain out of the solution. For this reason, you may wish to tape the alligator clip wire to the outside of the beaker to keep it from moving.
11. Attach the alligator clip of the second wire to the exposed end of the copper wire electrode hanging outside the beaker.
12. Record the initial solution level in the graduated cylinder, noting that the volume markings are inverted.
13. Carefully attach the free end of the unknown metal electrode wire to the positive terminal of the battery power source (or positive lead of the wall power source) using the alligator clip.
14. Carefully attach the free end of the copper electrode's wire to the negative terminal of the battery power source (or negative lead of the wall power source) using the alligator clip. You should immediately observe bubbles being produced inside the inverted graduated cylinder.

15. Notice that the H_2 gas produced rises into the graduated cylinder, displacing the solution.
16. Operate the electrolytic cell until you've collected approximately 70 mL of gas. Then stop operation by either unplugging the wall source or disconnecting an alligator clip from a battery terminal.
17. After all bubbles have risen to the solution surface, record in your lab manual the final solution level in the graduated cylinder, noting that the volume markings are inverted.
18. Measure the solution temperature in the beaker and record it in your lab manual. We can assume that the temperature of the solution equals the collected gas temperature.
19. Ask your instructor for today's measured atmospheric pressure. Record this value in your lab manual (see note below).

    NOTE: In this experiment, it is not accurate enough to assume that the atmospheric pressure equals the pressure of the collected H_2 gas in the graduated cylinder. The solution vapor pressure inside the graduated cylinder (which is essentially the same as the water vapor pressure) also contributes significantly. In other words,

    $$P_{\mathrm{H_2}} + P_{\mathrm{H_2O}\,vap} = P_{atm}$$

    Therefore, the vapor pressure of water at the solution temperature must be subtracted from atmospheric pressure to determine the H_2 gas pressure.
20. After disconnecting all wires carefully, remove the metal electrode from the solution and rinse it with distilled or deionized water.
21. Dry and clean the electrode thoroughly. Using the same balance you used before, record the mass of the electrode.
    After disposing of the remaining solution according to your instructor's directions, repeat steps 1-21 to perform a second trial.
23. When you are finished, dispose of all waste in the appropriate waste container. Return the clean electrode to your instructor.

Experiment 22 Pre-Lab (Show ALL Work and ALWAYS include units)

The experiment described above is performed with an unknown metal electrode, and the following data are obtained:

Mass of metal electrode prior to use in cell: 3.294 g
Initial solution level in the inverted graduated cylinder: 2.3 mL

Final solution level in the inverted graduated cylinder: 72.5 mL
Mass of metal electrode after use in cell: 3.228 g
Temperature of solution: 21.0 °C
Atmospheric pressure at time of experiment: 730.0 mm Hg

Use these data to answer questions 1–4.

1. In this experiment, it is not accurate enough to assume that the atmospheric pressure equals the H_2 gas pressure. The solution vapor pressure inside the graduated cylinder (which is essentially the same as the water vapor pressure) also contributes significantly. In other words,

$$P_{H_2} + P_{H_2O\,vap} = P_{atm}$$

Therefore, the vapor pressure of water at the solution temperature must be subtracted from atmospheric pressure to determine the H_2 gas pressure. Look up a value for the vapor pressure of water at the given solution temperature, record this value and your source, and then determine the pressure of the H_2 gas in mm Hg. For your information 760 mm Hg = 1 atm.

2. Calculate the moles of H_2 gas collected (use $R = 0.08206\ \frac{L*atm}{mol*K}$).

3. Calculate the moles of electrons transferred while the cell was in operation.

4. Calculate the molar mass of the metal and, using a periodic table, give its identity.

Experiment 22 Lab Report (Show ALL Work and ALWAYS include units)

Unknown # of electrode: ________

	Trial 1	Trial 2
1. Initial mass of clean electrode (g):	_____	_____
2. Initial solution level in cylinder (mL):	_____	_____
3. Final solution level in cylinder (mL):	_____	_____
4. Temperature of solution (°C):	_____	_____
5. Atmospheric pressure (mm Hg or torr):	_____	_____
6. Final mass of electrode (g):	_____	_____
7. Volume of H_2 gas collected (mL):	_____	_____
8. Mass loss of metal electrode (g):	_____	_____

Look up the vapor pressure of water at the same temperature as your solution. Report that value here and cite your source.

9. Vapor pressure of water at solution temp: _____

Source:

10. Calculate the H_2 gas pressure, being sure to account for the vapor pressure of the solution as indicated in the procedure.

11. Using the ideal gas law, calculate the moles of H_2 gas collected for each trial (use $R = 0.08206\ \frac{\text{L*atm}}{\text{mol*K}}$).

Trial 1:

Trial 2:

12. Calculate the moles of electrons transferred while the cell was in operation for each trial.

Trial 1:

Trial 2:

13. Calculate the moles of metal oxidized while the cell was in operation for each trial.

Trial 1:

Trial 2:

14. Calculate the molar mass of the metal based on your results for each trial.

Trial 1:

Trial 2:

15. Average molar mass of the metal from trials 1 and 2 (g/mol): ______

16. Identity of the unknown metal: ____________________

Experiment 23
Synthesis of $[Co(en)_2Cl_2]Cl$ *cis* and *trans* Isomers – Does Bond Placement on a Compound Matter?

Objectives
- Learn some inorganic synthesis techniques.
- Recognize that coordination compounds can have geometric isomers.
- See the effect of geometric isomerism on compound color.

Topics Introduced in this Lab
- Coordination compounds
- Ligands
- Geometric isomers
- Nomenclature

Related Zumdahl Sections
- Chapter 20 Sections 3 and 4

Safety Considerations
- 12 *M* and 6 *M* HCl(*aq*) are both strong and concentrated acid solutions that can cause severe skin burns. HCl in eyes results in serious eye damage. Be very careful with this substance. If any gets on your skin or eyes wash with copious amounts of water.
- 15 *M* $NH_3(aq)$ is a concentrated basic solution that can cause skin burns and has a noxious odor. Be very careful with this substance. If any gets on your skin or eyes wash with copious amounts of water.
- Cobalt compounds can cause some respiratory effects and skin allergies. Handle the cobalt(II) chloride with care. Dispose of this and all cobalt compounds in an approved waste container. Do not pour it down the drain.
- Boiling water spilled onto skin can cause serious burns. Handle the beaker of boiling water with care.
- Do not ingest any of these compounds.
- Wear safety glasses and appropriate lab wear at all times while in the lab.

Introduction
d-orbital transition metals form coordination compounds in which a compound or ion bonds to the metal's *d*-orbitals. Neutral compounds bound to a transition metal are called ligands. Ligands can bind to one, two, three or sometimes more of the transition metal's *d*-orbitals. For example, carbon monoxide (CO) has one lone pair of electrons on the C atom. In the presence of metallic Fe, carbon monoxide reacts to form trigonal bipyramidal shaped $Fe(CO)_5$, see **Figure 1**. Likewise, Co^{3+} ions bind to the lone pair of electrons on the N atom in ammonia to form an octahedral shaped ion, $[Co(NH_3)_6]^{3+}$. Both CO and NH_3 are called monodentate ligands, indicating that they bond to one place on the *d*-transition metal.

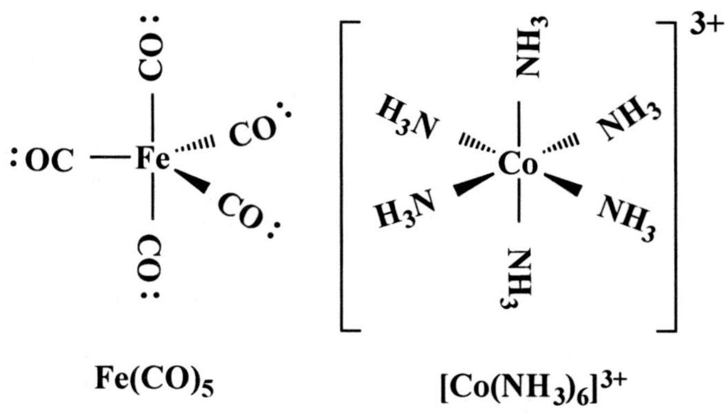

Figure 1. Fe and Co(III) Coordination Compounds

There are also bidentate, tridentate, etc. ligands that bond to *d*-transition metals in two, three, or more places. For example, oxalate ion, $C_2O_4^{2-}$, is a bidentate ligand. Diethylenetriamine, $C_4N_3H_{12}$, is a tridentate ligand. Coordination compounds containing these two ligands are shown in **Figure 2.**

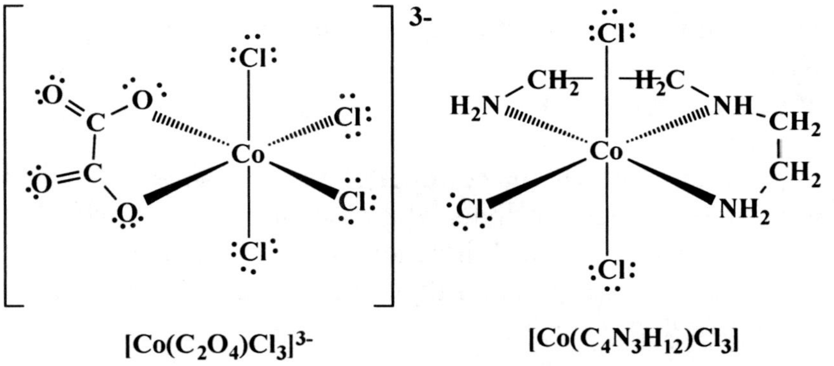

Figure 2. Bidentate and Tridentate Ligand Co(III) Coordination Compounds

Coordination compounds have several common molecular shapes, including square planar, trigonal bipyramidal, and octahedral. All three shapes can exhibit a form of isomerism called geometric or *cis, trans* isomerism. In geometric isomers the *cis* isomer occurs when two ligands of the same type are adjacent to each other. *Trans* isomers occur when two ligands of the same type are directly across from one another. In **Figure 3** you can see examples of *cis* and *trans* isomers for a square planar and an octahedral complex.

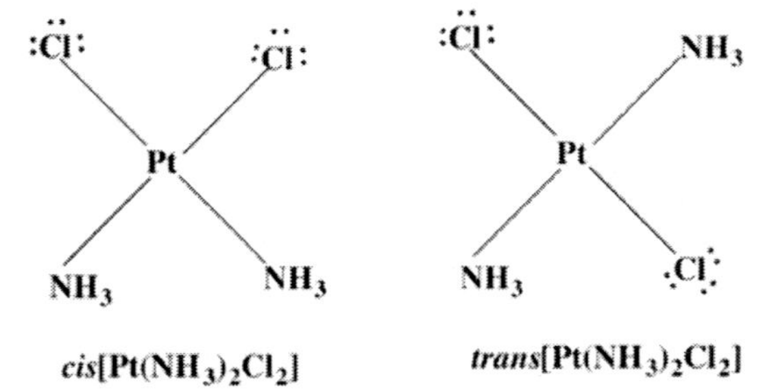

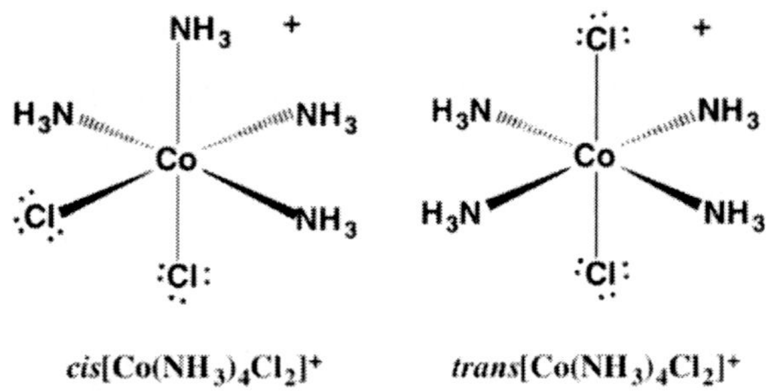

Figure 3. Geometric Isomers of $[Pt(NH_3)_2Cl_2]$ and $[Co(NH_3)_4Cl_2]^+$. To Visualize *cis*- and *trans*- Isomers in Octahedral Complexes Note the Cl^- Ion Positions in $[Co(NH_3)_4Cl_2]^+$.

In this experiment you will make both geometric isomers of an octahedral Co^{3+} complex containing two units of ethylenediamine and three chloride ions having the molecular formula $[Co(en)_2Cl_2]Cl$. Ethylenediamine, $C_2N_2H_8$, is a neutral, bidentate ligand frequently abbreviated as en. Ethylenediamine binds to *d*-transition metals as indicated in **Figure 4.**

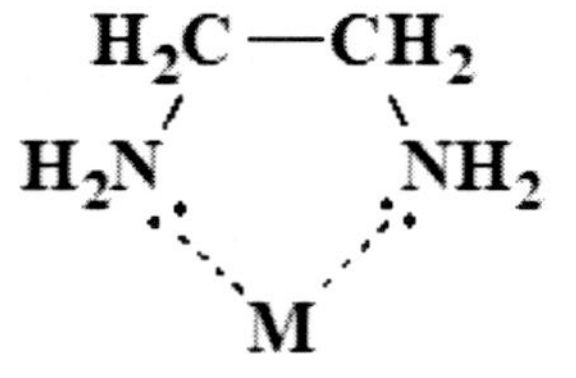

M represents any *d*-transition metal or ion

Figure 4. Ethylendiamine (en) Bonding Arrangement to a Metal Atom or Ion

Co^{2+} is much more reactive than Co^{3+}. In the presence of air, Co^{2+} is easily oxidized to Co^{3+}. We will take advantage of this fact by starting with $CoCl_2$ as the source of Co^{2+}. As the Co^{2+} reacts with ethylenediamine and excess chloride ions, frequently stirring the mixture allows air oxidation to occur, resulting in $[Co(en)_2Cl_2)Cl$ which contains Co^{3+}.

Procedure

Part A – Synthesis of *trans*-[Co(en)$_2$Cl$_2$]Cl

1. Obtain 4.0 g of $CoCl_2 \cdot 6\ H_2O$, 15 mL of 10% by mass ethylenediamine (*aq*) solution, an evaporating dish, and a 400-mL beaker.
2. Place all 4.0 g of $CoCl_2 \cdot 6\ H_2O$ and the 15 mL of ethylenediammine solution into the evaporating dish. Add 10 mL of distilled or deionized water and stir to dissolve the crystals.
3. Add 200 to 300 mL of water to the 400-mL beaker. Place the beaker onto a heating plate, then bring the water to a moderate boil.
4. Place the evaporating dish containing dissolved $CoCl_2 \cdot 6\ H_2O$ crystals on the top of the heated 400-mL beaker. This arrangement provides a constant steam source to drive the reaction.
5. Slowly stir the mixture using a stirring rod. As the mixture volume decreases add distilled or deionized water. Because stirring introduces air into the solution resulting in oxidation of Co^{2+} to Co^{3+}, it is important to stir constantly for at least 40 minutes. Move the experiment to a fume hood for the next step.
6. After 40 minutes of constant stirring, add 12 mL of concentrated HCl to the evaporating dish. Continue to heat and stir but do not add any more water to the mixture.
7. Over time small crystals of *trans*-[Co(en)$_2$Cl$_2$]Cl will appear as a slurry in the evaporating dish.
8. Using tongs or a heat resistant pad, remove the evaporating dish from the beaker top and place it into an ice-water bath. Allow the mixture to cool in the ice-water bath for at least 15 minutes.
9. Filter the mixture using either a funnel and filter paper or a suction filtration flask as directed by your instructor.
10. Move to a fume hood or other fume extraction device for the next three steps.
11. While still on the filter, wash the *trans*-[Co(en)$_2$Cl$_2$]Cl crystals twice with 6 *M* HCl(*aq*).
12. Scrape the crystals from the filter paper into a clean dry evaporating dish.
13. As directed by your instructor, either allow the crystals to air dry over night or heat the crystals over a steam bath or underneath a heat lamp. Break up any crystal clumps with a spatula or stirring rod to allow excess HCl and water to evaporate.
14. Collect your *trans*-[Co(en)$_2$Cl$_2$]Cl crystals. Record the color of the *trans*-[Co(en)$_2$Cl$_2$]Cl crystals.

Part B – Synthesis of *cis*-[Co(en)$_2$Cl$_2$]Cl

15. Place approximately 0.3 g of the dried *trans*-[Co(en)$_2$Cl$_2$]Cl crystals into a clean and dry evaporating dish.
16. Dissolve the *trans*-[Co(en)$_2$Cl$_2$]Cl crystals in the minimum amount of water necessary.
17. Add 1 drop of 15 *M* NH_3(*aq*) to the dissolved *trans*-[Co(en)$_2$Cl$_2$]Cl crystals. If any of the colored *trans*-[Co(en)$_2$Cl$_2$]Cl crystals remain, add one more drop of 15 *M* NH_3(*aq*).

18. Heat the evaporating dish over the 400-mL beaker containing water at a moderate boil. Continue heating until the salt changes color and the crystals are dry. Record the color of the *cis*-$[Co(en)_2Cl_2]Cl$ crystals.

Experiment 23 Pre-Lab (Show ALL Work and ALWAYS include units)

1. The compound you will make in this lab has the molecular formula $[Co(en)_2Cl_2]Cl$. Explain why two chloride ions are placed inside the brackets and one outside. This requires a structural explanation. It may be necessary to draw the structure as a part of your explanation.

2. What is the ionic charge of Cr in the compound $[Cr(en)_2Cl_2]Cl_2$. Explain how you came to this conclusion.

3. Draw the *cis* isomer of $[Co(en)_2Cl_2]Cl$.

4. Draw the *trans* isomer of $[Co(en)_2Cl_2]Cl$.

Experiment 23 Lab Report (Show ALL Work and ALWAYS include units)

Part A – Synthesis of *trans*-$[Co(en)_2Cl_2]Cl$

1. What color is $CoCl_2 \cdot 6\ H_2O$? ________

2. What color is *trans*-$[Co(en)_2Cl_2]Cl$? ________

Part B – Synthesis of *cis*-$[Co(en)_2Cl_2]Cl$

3. What color is *cis*-$[Co(en)_2Cl_2]Cl$? ________

4. Why must you stir the $CoCl_2 \cdot 6\ H_2O$ and HCl mixture so much to make the reaction occur?

5. Why is the transformation of *trans*-$[Co(en)_2Cl_2]Cl$ to *cis*-$[Co(en)_2Cl_2]Cl$ so much simpler and faster than the first reaction?

6. What noticeable effect did the switch from *trans* to *cis* isomers have on $[Co(en)_2Cl_2]Cl$?

Experiment 24
Using Distillation to Purify Solvent

Objective
- Use distillation to purify an organic solvent.

Topics Introduced in this Lab
- Distillation
- Alcohols

Related Zumdahl Sections
- Review Chapter Section 8
- Chapter 21 Section 4

Safety Considerations
- The setup in this experiment gets very hot. Use caution.
- Ethanol is flammable. No open flames should be used in the lab during this experiment.
- Do not ingest any of these compounds.
- Wear safety glasses and appropriate lab wear at all times while in the lab.

Introduction
Ethanol is an organic molecule called an alcohol. Alcohols are easily distinguished by the presence of an –OH group attached to a hydrocarbon chain. Specifically, ethanol has the structural formula CH_3CH_2OH, shown with the individual bonds in **Figure 1**. Other common alcohols include methanol and isopropyl alcohol (also called isopropanol), whose structures are also shown in **Figure 1**. Methanol is highly toxic and can cause blindness upon ingestion. Isopropanol is found in many households as a solution in water and has the common name "rubbing alcohol". All three of these alcohols can be used as reagents or solvents in organic chemistry.

Organic chemists often need to purify solvents or liquid reagents that they will then use in a synthetic reaction. These solvents can be purified using a technique known as distillation. This technique relies on different boiling point temperatures of the various liquids in a mixture. Those liquids with lower boiling points can be evaporated and re-condensed into separate vessels, leaving the other, higher boiling point liquid "contaminants" behind. In this experiment, you will perform a distillation to purify ethanol from a mixture of ethanol, water, and food coloring.

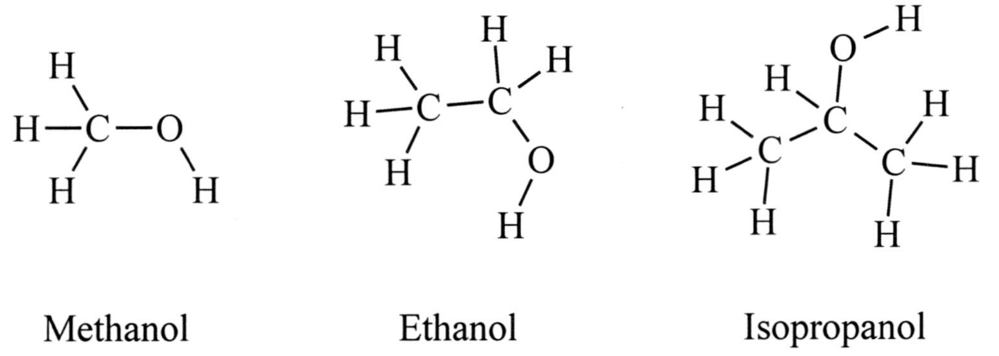

Figure 1. Structure of Methanol, Ethanol, and Isopropanol

Procedure

1. Set up your distillation apparatus as shown in **Figure 2**.
    a. Use joint grease to thoroughly grease all of the glassware connections before assembling your apparatus.
    b. Be sure to use two utility clamps as indicated. (These clamps will be attached to two different rings stands, one for each clamp.)
    c. Connect the rubber tubing from the outermost water inlet port of the condenser tube to a water faucet.
    d. Connect a second piece of rubber tubing to the other water port on the condenser tube, then run it into a laboratory sink.
    e. When the water is turned on, make sure the flow occurs in the direction indicated in **Figure 2**.
    f. The tip of the thermometer should be located at the point of the distillation neck where the junction occurs.
2. After setting up your apparatus, obtain approximately 100 mL of the ethanol/water/food coloring mixture and add it to your 250-mL roundbottom flask. (You will have to detach the flask to fill it, and then re-attach it afterwards.)
3. Add a few boiling stones to the roundbottom flask. (NOTE: Boiling stones should never be added to a solution that is already hot)
4. Check to make sure the cavity of the thermowell you will be using is filled with sand. Plug the thermowell into a variac, and make sure the variac is not turned on. Then plug the variac into an electrical outlet.
5. Position the distillation apparatus so that the roundbottom flask is seated in the sand within the thermowell. Be sure that all joints on the apparatus are tightly sealed.
6. Position a 50-mL beaker beneath the open end of the condenser tube. This will serve to collect the purified ethanol.

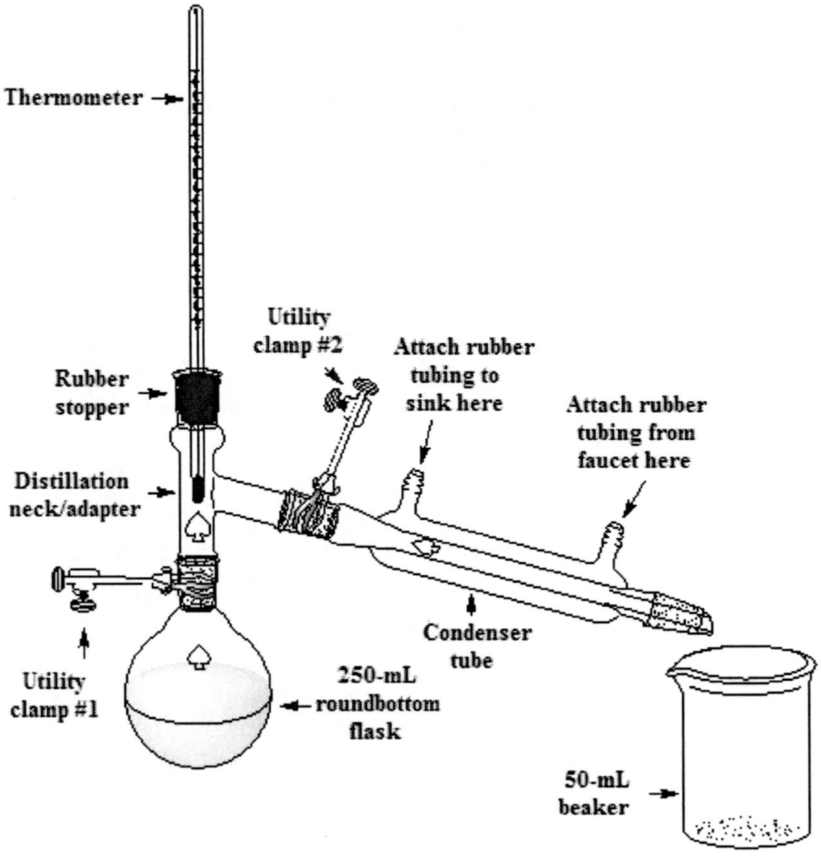

Figure 2. Distillation Apparatus

7. Use aluminum foil to completely cover the outside of the roundbottom flask and distillation neck all the way up to the start of the condenser tube. Do not put any foil between the sand and roundbottom flask. The aluminum foil minimizes heat loss during the distillation.
8. Turn on the variac to 100%. Be careful, as the thermowell, sand, and distillation apparatus will get very hot.
9. Monitor the temperature reading on the thermometer very carefully.
10. Once the temperature begins to rise, quickly turn the variac back to 0% and let the temperature stabilize.
11. From this point, manipulate the variac to reach and maintain a temperature of roughly 80° C.
12. After reaching the desired temperature, you should begin to observe liquid condensing in the condenser tube and dripping into your beaker. What liquid is this? How do you know? What color is it and what does this tell you about the food coloring?
13. After you have collected at least 15 mL of distilled liquid, use a 10-mL volumetric pipet to determine the density of the liquid. What is the liquid's density? Does this confirm the identity of the liquid?
14. Turn off the variac and allow your apparatus and remaining solution to cool. Once cool, pour the remaining solution down the sink. Clean all glassware.

Experiment 24 Pre-Lab (Show ALL Work and ALWAYS include units)

1. Look up the normal boiling points of ethanol and report the values here. Be sure to cite your source(s).

   Normal Boiling Point of Ethanol: ______

   Normal Boiling Point of Water: ______

   Source(s):

2. If you performed a distillation of a mixture of water and ethanol, heating the mixture to approximately 80 °C, which liquid would you expect to collect from the condenser? Explain.

3. If while performing a distillation of ethanol and water, you heat the mixture above 100 °C, what would you expect to happen?

Experiment 24 Lab Report (Show ALL Work and ALWAYS include units)

1. Look up the boiling points of ethanol and water at sea level and report the values here. Be sure to cite your source(s).

   Boiling Point of Ethanol (at sea level): ______

   Boiling Point of Water (at sea level): ______

   Source(s):

2. While performing the distillation, what liquid condensed in the condenser tube and dripped into your beaker? How do you know?

3. What color was the liquid you collected from the condenser tube and what does this tell you about the food coloring?

4. Calculate the density of the liquid collected from the condenser. Does the calculated density confirm the identity of the liquid? If yes, explain. If no, why does the density not match that of ethanol, water, or food coloring?

Experiment 25
Chemical Spill – A Self-Directed Experiment

A local chemical supply company was transporting a large shipment of solid copper(II) sulfate pentahydrate ($CuSO_4 \cdot 5H_2O$). The transport truck was involved in an accident, and while there were no serious injuries, thousands of kilograms of copper(II) sulfate pentahydrate spilled directly into a small lake beside the road. As an analytical chemist for the state, you have been brought in to determine the exact concentration of Cu^{2+} in the water, so that any dangers to the community and the local aquatic life can be determined.

In this self-directed experiment you will have access to a sample of water from the lake, some of the solid copper(II) sulfate pentahydrate that the truck was carrying, all common laboratory supplies and glassware, a 10-mL volumetric flask, as well as all of the equipment needed for absorption spectroscopy. Based on estimates of the amount of water in the lake and the amount of chemical the truck was carrying, you know that the concentration of Cu^{2+} in the lake cannot be more than 0.1 *M* or less than 0.01 *M*. You are to determine the exact concentration of Cu^{2+} in the lake water and report your findings to the state and local governments (in this case your instructor).

Prior to the experiment, you must write your proposed procedure for this experiment and present it to your instructor. He or she will read your proposed procedure, make suggestions on the proposal while ensuring that the procedure is safe, and then return it to you prior to the experiment. You are expected to perform the procedure on your own without assistance from other students or the instructor. After you perform the experiment, you will write a lab report which describes the procedure you used, the results you obtained, and the conclusions you can draw from your results. Ask your instructor about any specific format guidelines for writing the proposed procedure and/or the lab report.

CPSIA information can be obtained
at www.ICGtesting.com
Printed in the USA
FFOW04n0655020916
27318FF